This book is dedicated to my parents

John and Verity Morrison

Never to be forgotten

STRUCTURALLY INSULATED PANELS DESIGN

Structural Design Manual

www.struartapp.com

First Edition 2024

ISBN 978-1-917096-00-3

Cover Design by David Colon

Printed and Bound by Amazon

STRUCTURALLY INSULATED PANELS DESIGN

Structural Design Manual

ACKNOWLEDGEMENTS

As much as this book was enjoyable to write, it would not have been possible without the endless patience of my wife Bernadette who has had to listen to my daily ramblings about its creation and helped with the proof reading. I am indebted to my former colleague from Anthony Hunt Associates days, Jon Carr, who proof read the book with his sharp eyes, skills and knowledge gleaned from his professional career and teaching at Sheffield University.

My loyal clients at SIP Co, Bentley Projects, Glosford and SIP Hus, together with their designers have provided me with years of interesting projects, which have been a source for much of the material for the book. With their enthusiasm for SIPs construction it has been possible to bring to fruition many difficult structures not normally associated with SIPs and hopefully to the satisfaction of clients.

Not forgetting my friends Dave Moss and Tracey Johnstone who have been very helpful and encouraging despite my boring dialogues over the dinner table and on the golf course.

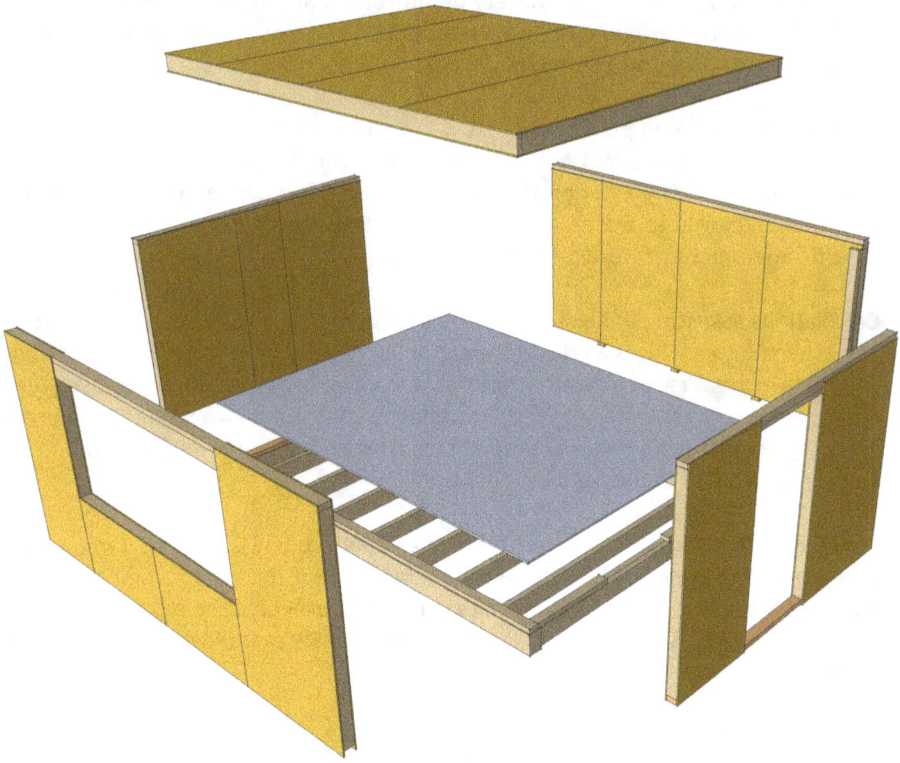

STRUCTURALLY INSULATED PANELS DESIGN

Structural Design Manual

DISCLAIMER

This book is intended for professionals familiar with structural design, particularly with knowledge of calculation methods and the ability to understand technical drawings.

A great deal of effort has been taken to ensure that the results presented by the calculations are correct. However, it cannot be guaranteed that this will be the case in all situations. It is the responsibility of any persons using this material to ensure that they become familiar with the methods and are confident that any calculations undertaken are giving reasonable results. It is strongly recommended to consult manufacturers' data, structural certification and standard details relevant to the specific SIPs panels intended for use, in order that the calculations and details may be checked for consistency with that product, particularly in relation to material properties, dimensions, fixings and connections.

All content herein was written on the basis of wide experience in the structural design of structurally insulated panel (SIP) construction. The best endeavours were made to combine material from a wide variety of international sources - manufacturers' data, academic papers, design guides and text books - in a manner that provides consistent and practical methodology based primarily on Eurocode limit state design.

It has not always been possible to follow Eurocodes in their entirety. They do not contain comprehensive material properties or cover all aspects of SIPs design. For instance, neither PUR, EPS nor XPS core engineering properties are provided, nor are they manufactured to a consistent standard in the industry. Designers must refer to specific manufacturers for verified properties.

Calculations assessing deflection of panels employ conservative core shear modulus and creep factors. This is a critical aspect of SIP panel construction in which engineers should take great care to ensure that they have valid data if they are to use greater core modulus values or lower creep (K_{def}) factors. Test data from panel compression tests and screw tension/shear in panel faces has been used to verify methods of analysis presented in the book.

For racking, adaptation of the former British Standard BS5268 Part 6.1 in Eurocode 5 (method B) is erroneous for dimensional factors. For this reason, among others, BS5268 Part 6.1 has been used on the basis that it proved to be a practical standard for over 25 years, is versatile and is consistent with SIP manufacturers' quoted racking strengths.

List of Notations

A_c Area of foam core
A_{cb} Effective area of core in bearing on a support wall
A_{eff} Effective area of foam core in flexure from Allen[2]
A_{pb} Effective area of core in bearing under point load
A_H Area of header in SIP wall as a beam
A_L Area of lintel under SIP wall as a beam
B_{eff} Effective width of reinforced panel (with *built in* joist) due to shear lag
B_j Width or breadth of timber spline (or *built in* joist)
B_p Panel width or breadth
B_s Width of stud
B_w Effective bearing width with respect to core compression
C Constant used to obtain natural frequency from deflection
C_v Reduction factor for shear resistance of panel fixed onto a side wall
D Flexural rigidity of panel
D_b Effective core diameter in bearing under a screw head from uplift
d_c Depth of SIP Core
d_{eff} Effective depth in flexure (*depth between face centres when* $t_1 = t_2$)
d_{fh} Diameter of flange headed screw
d_j Depth of timber spline (or *built in* joist)
d_{max} Maximum deflection (at centre span for panel with perpendicular UDL)
$d_{max(Gk)B}$ Maximum deflection/sag due to dead load from face bending component
$d_{max(Gk)V}$ Maximum deflection/sag due to dead load from core shear component
$d_{max(Qk)B}$ Maximum deflection/sag due to live load from face bending component
$d_{max(Qk)V}$ Maximum deflection/sag due to live load from core shear component
d_p Depth of SIP panel
d_s Depth of stud within wall
e Eccentricity
$E_{f(0.05)}$ *Fifth percentile* characteristic face or stud modulus of elasticity
$E_{(mean)}$ Mean characteristic value of face modulus of elasticity parallel to grain
E_f Young's modulus of SIP panel face
E_j Young's modulus of timber spline (or *built in* joist)
E_p Idealised effective Young's modulus of panel
$E_{x,y,z}$ Modulus for x,y,z axes for plate or shell element
F Applied racking load (unfactored) or *point load*
$F_{ax,d}$ Design screw tension (or withdrawal) resistance
$F_{ax,rk}$ Characteristic screw tension (or withdrawal) resistance
$f_{c,0,k}$ Characteristic resistance in compression of face
$f_{cc,PUR/EPS}$ Compressive Strength (characteristic compressive resistance) of core
$F_{cd,PUR}$ Design compressive resistance of core
$F_{c,0,d}$ Applied design compression on wall
$F_{cd,PUR/EPS}$ Design Compressive resistance of core
$f_{head,k}$ Characteristic pull through resistance of screw head through material face
F_H Header force in SIP wall with beam action
$f_{h,0,d}$ Design shear capacity of bolt fixing
F_L Lintel force in SIP wall with beam action

List of Notations (continued)

$f_{m,d}$	Design stud bending strength
$f_{m,k}$	Characteristic stud bending strength
F_n	Fundamental natural frequency of floor
F_{ne}	Racking load in wall calculated from its locality in a system
$F_{p,v,d}$	Permanent design resistance of wall fixings (shear)
F_{sv}	Total Resistance to Sliding of SIP or stud wall
$F_{t,d}$	Instantaneous design suction resistance of wall fixings (tension)
$f_{t,0,k}$	Characteristic resistance in tension of face
F_{vc}	Characteristic shear resistance of panel with SIP splines
$F_{vc,rein}$	Characteristic shear resistance of panel with timber splines
F_{vd}	Design shear resistance of panel with SIP splines
$F_{vd,rein}$	Design shear resistance of panel with timber splines
$F_{v,d}$ ($F_{v,r}$)	Design screw/bolt/anchor fixing shear resistance ($F_{v,r}$ - soleplate anchor)
$F_{v,r}$	Recommended Shear resistance of wall soleplate anchors
$F_{v,rd}$	Design shear resistance of wall soleplate anchors
$F_{v,rk}$	Characteristic screw shear resistance
G_c	Shear modulus of foam core obtained from manufacturers' data
$G_{c(B)}$	Flexural shear modulus of foam core derived from shear modulus, G_c
G_k	Dead Load (UDL on panel)
G_{ks}	Dead load perpendicular to sloping panel
h	Racking SIP or stud wall height
I_f	Moment of inertia of SIP faces ($I_f = (dp^3 - dc^3).Bp/12$)
I_p	Idealised effective moment of inertia of panel
$K_{def,f}$	Load duration factor for SIP panel faces
$K_{def,c}$	Load duration factor for SIP panel core
$K_{def,t}$	Load duration factor for timber joist or stud
$k_{c,y}$	Buckling factor for stud (Y axis)
K_e	Eccentricity coefficient based on secant formula
k_m	Strut restraint coefficient to obtain L_{eff}
K_{mod}	Modification factor for load duration
$K_{mod,j}$	Modification factor for load duration for timber spline (or *built in* joist)
K_n	Stiffness of racking wall (assumed proportional to length of wall)
k_{sys}	System coefficient for multiple studs in a wall
K101 *etc.*.	Modification factors for racking SIP or stud walls
L	Span of beam or floor to floor height of post/column or racking wall length
L_{eff}	Effective height of post depending on end restraints
Ln	Length of wall which is part of a group resisting racking loads
L_s	Slope length of panel or beam (L/cosθ)
$M_{d,c}$	Design bending resistance in compression
$M_{d,t}$	Design bending resistance in tension
$M_{ecc,Gk/Qk}$	Eccentric moment in wall at mid height from dead/live load
M_{OT}	Overturning moment of racking SIP or stud wall
M_r	Modular ratio
M_R	Restoring moment of racking SIP or stud wall
$M_{r,d}$	Design moment resistance of SIP panel or wall

List of Notations (continued)

M_x	Moment distribution in plate or shell finite element model about X axis
M_y	Moment distribution in plate or shell finite element model about Y axis
$M_{y,0,d}$	Applied design transverse moment in SIP wall
N_{xy}	In-plane or racking shear in finite element model plate or shell elements
P_b	Characteristic bearing resistance of wall in compression
P_{bs}	Characteristic compression resistance of stud
$P_{b,d}$	Design bearing resistance of wall in compression
$P_{c,o,d}$	Design axial compression resistance of SIP wall
P_{cr}	Critical SIP wall buckling load
P_{cre}	Critical SIP *eccentrically loaded* wall buckling load
$P_{cre,d}$	Design SIP *eccentrically loaded* wall buckling load
$P_{cr,s}$	Critical buckling of stud in SIP wall
$P_{cr,s,d}$	Design buckling of stud in SIP wall
$P_{d,PT}$	Design pull through resistance of panel face at screw head under uplift
P_E	Euler critical buckling load of wall in compression
$P_{ld,PUR/EPS}$	Design compression resistance of panel as a line load (*eg supported on wall*)
$P_{pd,PUR/EPS}$	Design compression resistance of panel supporting a point load
$P_{w,o,d(600)}$	Design resistance of stud wall in compression (studs at 600mm centres)
Q_{ks}	Live load perpendicular to sloping panel
R_{dv}	Design racking resistance of a SIP or stud wall
Rfrv	Friction resistance of SIP or stud wall, serviceability state
R_{rk}	Racking resistance of SIP or stud wall calculated from Rv, length and K' factors
Rv	Unfactored, or recommended, racking resistance of a SIP or stud wall
r_y	Radius of gyration of a SIP panel
S_f	Spacing of fixings or soleplate anchors
S_s	Stud spacing
t_1, t_2	Thickness of faces forming SIP panel
UDL	Uniformly distributed load
V	Shear in panel
V_d	Design shear force at panel support
V_{ph}	Shear flow or rolling shear in header of SIP wall as a beam
Y_{na}	Neutral axis position in SIP wall as a beam
w	Applied self weight UDL on racking wall
Ws	Perpendicular load on sloping panel (W.cosθ)
Wu	Ultimate perpendicular load on panel
Zn	Distance of a wall from the centroid of a group resisting racking loads
β_c	Stud straightness factor
Υm	Partial factor for material property
$\sigma_{m,crit}$	Critical stress due to moment in stud post
η_{eff}	Effective number of bolts in a group
θ	Panel or beam slope with reference to horizontal
$\lambda_{rel,m}$	Relative slenderness of stud subjected to bending
$\lambda_{rel,y}$	Relative slenderness of stud (Y axis)
λ_y	Slenderness of stud (Y axis)
μ	Friction coefficient of SIP or stud wall at the base support
$\psi_{1,2}$	Quasi-permanent factor (ψ_2) for variable action (*live load*)

STRUCTURALLY INSULATED PANELS DESIGN

Structural Design Manual

CONTENTS

Chapter 1

Introduction to SIPs

Background. Sustainability and energy criteria. Examples of SIPs buildings Typical structural assemblies and details. SIPs in construction. Summary of book. Resources and references.

1.1 INTRODUCTION TO SIPS

Structurally Insulated Panels (SIPs) have become firmly established in the construction of houses, hotels, leisure centres, garden offices and a host of other building types over the last 35 years in the UK and since the 1970s in USA. They were originally developed by the architect Alden B. Dow, with polystyrene panels (1.-5/8" thick or 41mm) with 1/4" (6mm) fir plywood boards. Foam was manufactured by his family company Dow Chemicals.

Modern SIPs panels may be fabricated from a variety of face skins, commonly: Orientated Strand Board (OSB), Magnesium Oxide Board (MgO), steel and aluminium. The insulated core may be bonded to the facing material - as in the case of Expanded Polystyrene (EPS) , also in extruded form (XPS) - or may be moulded between the two facings under pressure; a common example being Polyurethane (PUR). Compressed straw has also been employed as a recycled material.

SIPs panels may be adopted in a great variety of structures. They offer on site assembly supplied in kit form; cut to size from accurate three dimensional drawings. The components or even whole structures may be assembled offsite and transported to site, offering all the benefits of prefabrication with rapid construction and reduced site activities; with time and cost saving benefits over traditional construction methods or systems.

SIPs exhibit excellent thermal performance and lightweight structures. They may be used in floors, walls and roofs for residential, commercial and educational applications, providing ecologically friendly and energy efficient buildings.

Figure 1.1.1 SIPs House Construction Site Episkopy Cyprus

SIPs Sustainability and Energy Criteria

According to the UK Energy Research Centre: "buildings in the EU account for approximately 40% of the final energy consumption and 36% of greenhouse gas emissions of the trading bloc." As a proportion of that, residential buildings account for around 77% of that energy use, or roughly 28% of total energy use. SIP construction has scope for considerable reduction in consumption of energy.

SIP buildings, as airtight and highly insulated construction, combined with controlled heating and ventilating systems, meet the criteria for significant energy savings over traditional methods of construction, at all stages from design leading to occupation.

The US Structurally Insulated Panels Association (SIPA) highlight economies in manufacturing. OSB facings are made from wood chips sourced from 90% of logs of inexpensive species. Highly automated machinery is employed in OSB production, with all waste products convertible to wood chips for fuel and bark dust for instance. Polyurethane and EPS insulation takes less energy to produce than fibreglass insulation, is a better insulator, has a life cycle of 50 years or more and has the potential for recycling at end of life.

As regards thermal insulation, Building Regulations in the UK currently have a target U value of 0.18 for walls, which is easily achieved with SIPs. As a comparison a wall constructed from a 150mm SIP panel, 50mm low emissivity air gap and 100mm brick face typically has a U value of circa 0.16. A cavity wall with inner leaf 100mm thermalite type block, 100mm cavity with low emissivity, 15mm air gap and outer leaf 105mm brick typically has a U value of circa 0.22.

Similarly SIP roof panels offer very good insulation and air tightness compared with traditional truss or rafter construction. The target UK current U value for roofs is 0.15. A SIP roof with battened top finishes and skimmed plasterboard ceiling typically meets this criterion, with additional insulation achieving circa 0.11.

Timber is generally employed throughout to trim large openings and voids, for trimming openings and for stud walls, purlins and ridge beams. Arguably this is the only 100% sustainable material, since trees may be replaced. Typically a timber framed domestic property uses 4 tonnes less embodied carbon than for an equivalent masonry alternative. Steel or engineered timber beams may be necessary to carry floors over large spaces or to transfer upper floors to ground, but that would be no different should the same layout be constructed in masonry.

Material savings and optimisation of manufactured elements, made possible with 3D computer aided design software, enables fast construction and savings in site labour. Typically a house structure may be erected in under three weeks onto a pre-built ground structure. This has great benefit compared with traditional construction, where site coordination frequently causes delays and impacts on affected trades.

There has been a growth in offsite structures, which may be factory built in modules, fitted out and assembled on site. As can be seen with the examples which follow SIPs systems have proven to be well adapted to the demands of modern construction.

1.2 PURPOSE OF THE BOOK

To date there are no known text books that offer comprehensive guidance for SIPs structural design. The intention is to collate design friendly methods of structural analysis that draw together strength and stiffness test data, where available, to provide a comprehensive guide to all aspects of design that an engineer will encounter in SIPs design.

To the knowledge of the writer, there are no specific design standards, aside from the USA, in Canada, UK, Europe or any other country. In the USA Eric Tompos, vice president of the National Timber Association, was instrumental is publishing a very useful and concise document: "Engineered Design of Structural Insulated Panels". This document may be easily found online (pdhonline.com: *https://tinyurl.com/2f2vnuxa*). It tackles the fundamentals of bending, shear strength, shear walls as diaphragms, bearing, compression as well as other design aspects. A more up to date standard: "Structural Insulated Panel Design Specification SIP-EDG01-19S" by SIPA was drafted in 2019. SIPs have been included in the International Residential Code (IRC) which recognises that SIP panels are equivalent to other code approved building systems (reference: *https//www.sips.org/resources/building-codes*).

In the UK engineers rely on available data and calculations derived largely from first principles for composite, stressed skin structures. Herein the basis for design checks, except where noted, is Eurocode 5: Recommendations for the Design of Timber Structures (EC5). An accompanying UK document, PD 6693-1, gives supplementary design guidance. British Standard for timber frame design BS5268 Part 6.1 is now officially obsolete, although it has been adopted for racking checks in **chapter 3**. Notwithstanding Eurocode 5, engineers must refer to the British Board of Agrément (BBA) certificates appropriate to the manufactured SIP panel specific to the project in hand for OSB facing properties, racking resistances and other parameters.

It is not a text book for the design of timber elements, such as studs in walls which need consideration for compression loading, timber lintels or steel frames. However, where the timber or steel element is an essential component of the structural integrity of the SIP wall, floor or roof assembly, structural properties have been derived from the design code Eurocode 5 and will be quoted where appropriate.

This book is largely based on limit state design to Eurocode 5 employing SI metric units. However, there is no reason why the methods cannot be adapted to permissible design methods by employing unfactored loading, compared with design resistances reduced in proportion with the aggregate ultimate load factors. For engineers who use, for instance, BS5268 Part 6.1 (Code of Practice for Timber Frame Walls) the design capacities calculated may be divided by an ultimate load factor. Appropriate advice will be given for all design criteria which follow.

The concluding section of this introductory chapter summarises the engineering design topics covered in the book. As a supplement, worked examples and matters of interest are intended to be added to a dedicated section in the author's website *www.struartapp.com*.

1.2.1 BOOK SUMMARY

Chapter 2: Bending (Flexure)

Analysis of plate or beam action of single and multiple panels supporting out of plane loads, together with associated aspects such as local point loading, is examined. SIPs panels are commonly used as roof structure in pitched and flat roofs in domestic and commercial construction. SIPs floors are employed in modular 'garden office' type buildings where the floor spans and beneficial insulative properties come into play.

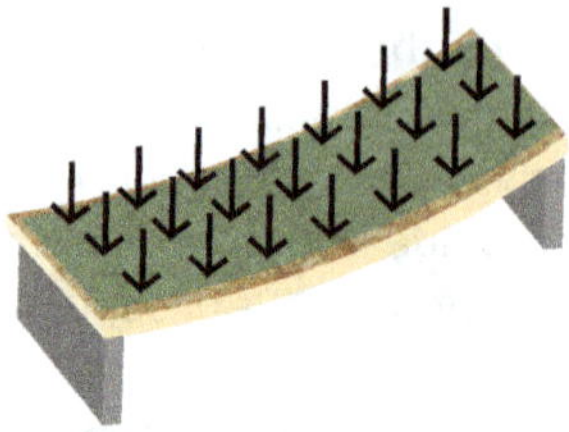

Bending (or Flexure)
(Sag/distortion under load)

Roofs generally have lower live loads from snow or occasional access. Of most concern is the phenomenon of long term sag, in which the 'creep' effect causes panels to sag further over a long period of time. In the diagram above an illustration of a natural green roof is shown; an application with high dead loads.

Chapter 3: Stability - Racking (In-Plane Shear)

Racking is essentially the resistance of a building to lateral loads: namely loading from wind or earthquakes, which cause horizontally applied (or lateral) forces to act on a building, which is stabilised by shear walls and additional framing if required. This chapter gives guidance on design checks for racking resistance of individual and combined stabilising structures.

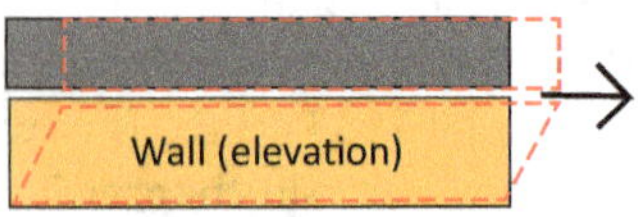

Racking
(Resisting lateral loads)

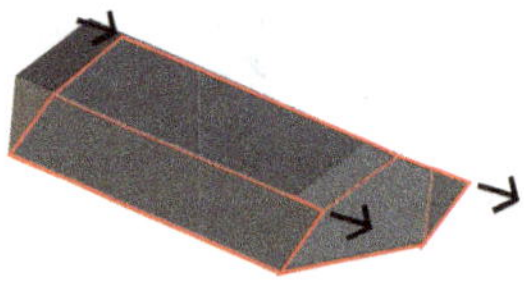

In the UK manufacturers publish the racking strength of panels based on standard testing; usually in BBA certificates. Published racking capacities vary significantly. This is dependent on test wall construction, particularly screw or nail fixings and adhesive or silicone sealants.

If a building has large openings then these may be strengthened with steel or timber framing to complement the overall racking resistance. Internal sheathed timber frame stud walls may also assist.

Chapter 4: Axial Compression (Walls)

This chapter provides guidance for the assessment of the compression resistance of walls carrying vertical loading, which may be from floor joists or roof panels.

For buildings up to three stories, supporting timber floors and roofs, this is not usually an issue providing the walls are no greater than circa 3m high between floors or roof. For taller or higher loaded walls the formulae presented are applicable.

Compression Loads

Where there are windows or door openings timber, or sometimes steel (if needed), is used to strengthen the edges and to provide solid fixing points for door or window framing. Lintels or shallow SIP panels provide support, as required, by spanning over the openings. As with roof and floor elements solid timber splines in joints strengthens SIP walls if heavier loads are to be supported. For example when precast concrete slabs act as floors.

There is a paucity of panel compression test data for use in the UK. Panels may be two stories (circa 5.5m) in height and vary in width. Euler derived buckling criteria are outlined, with consideration of the condition of eccentric loading. Calculated design compression resistance values are compared with published test data. Guidance to the assessment of timber stud walls in compression is included.

Chapter 5: SIPs as a Beam

Methods to appraise panels spanning across openings, from first principles, as composite beams, are presented. The panel assembly shown to the right is a floor height panel. By performing as a deep beam (similar to a diaphragm) they are able to support loads in bending and shear action.

Shallow SIPs panels in wall openings may also be employed as load bearing lintels.

Care must be taken to check the OSB/timber stresses and loading on fixings.

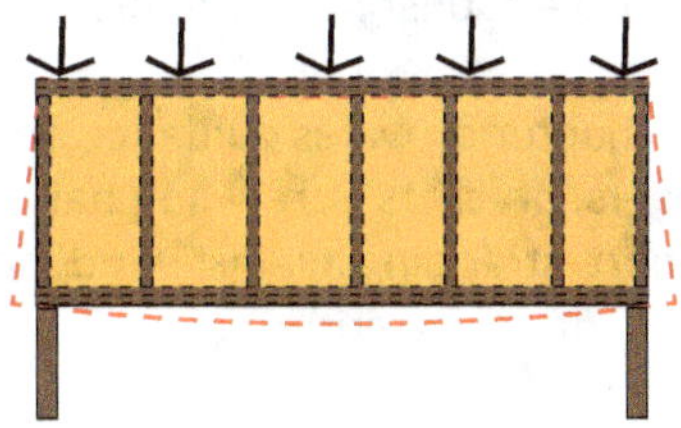

Load Bearing Wall (or Lintel)
(Spanning between supports)

Chapter 6: SIPs as Cladding Support

Guidance is given to assess SIPs wall and roof panels, supporting cladding; including render on renderboard, lapped boards, tiles and brick slips. Supporting screw shear and pullout test data is examined to obtain design resistances in OSB and Magnesium Oxide (MgO) faces, in conjunction with supporting battens or rails.

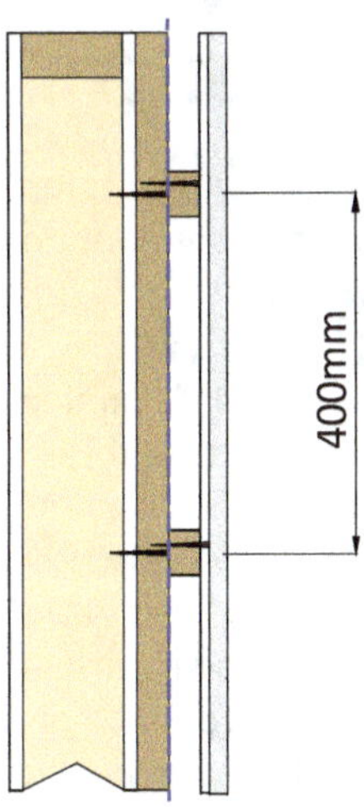

Masonry is generally too heavy to be carried by SIPs. Brick slips, which are thin brick elements fixed to a substrate board, may be supported in this manner owing to their lighter weight. **Section 1.5** shows a typical assembly with brick outer face, offset with a cavity that is commonly 50mm wide.

The chapter considers aspects of SIPs as a cladding support, their resistance to wind, issues of tolerance and relative movement with the adjacent building fabric.

Examples will be discussed together with design considerations; for instance when a brick skin needs to be carried over an opening.

When SIPs panels are fixed to multistorey structures greater than three stories, the effects of differential shrinkage with masonry faced buildings necessitates horizontal joints.

Chapter 7: Modular Construction

In this chapter SIPs modular units are reviewed, encompassing garden offices, site cabins, garden leisure rooms and chalet holiday accommodation, to name a few.

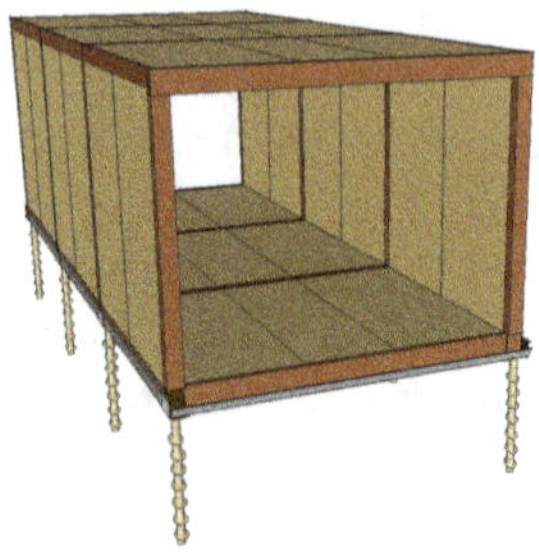

These may be offsite or 'on site' installation. Offsite assembly combines the benefits of clean assembly in a controlled environment and management of the workforce.

Fully fitted out and serviced units may be prefabricated and transported to site. Transport is a limiting factor. On UK roads the maximum width of a unit which may be carried is 2.9m without escort or police notice.

Chapter 8: Connection Details

Connections and fixings, which are frequently used in SIPs construction, are presented, which are designed to achieve robust installations.

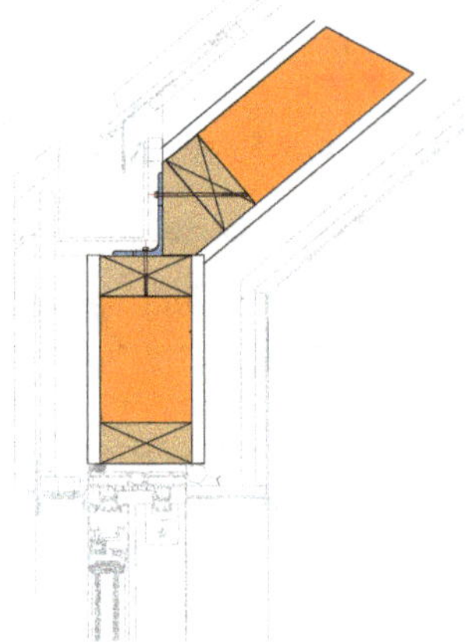

This chapter provides guidance on typical details, hangers, brackets, screw and nail fixings, joist hanger support, strapping against wind loads and other pertinent design criteria.

One of the problems encountered when trying to design to meet the requirements of codes and standards is a lack of easily sourced test data; for instance the compressive or crushing capacity of foam core or pull out resistances of screws and nails in the faces, whether OSB or MgO board. References to relevant chapters are made where appropriate.

Chapter 9 The Future of SIPs

This chapter puts forward proposals to improve certification and reviews unexplored and untested areas of modern construction which would encourage wider adoption of SIPs and to enable more economies in materials and construction.

For schools and larger commercial buildings cross laminated timber (CLT) is finding a niche. There is no reason why the insulative benefits of SIPS cannot be combined in hybrid structures. When SIP panels are seated on supporting beams, such as roof purlins, composite action as a potential to enhance strength and rigidity is discussed. Further testing, which could provide to engineers greater confidence in their designs, potentially achieving economy, is proposed. The use of adhesive bonding in timber frames, such as wind frames is discussed. Multistorey offsite erection would be of significant benefit. With ever greater demand for housing, fast, sustainable and high quality construction is possible.

1.3 EXAMPLES OF SIPS CONSTRUCTION

SIPs are suitable for many types of sustainable construction in domestic and commercial use. A brief summary of projects follows, including domestic, modular units, schools, commercial buildings and SIPs fixed to a structural frame, all of which the writer has been commissioned to design in the last fifteen years.

1.3.1 Domestic (House or Extension)

Probably the most common kind of SIPs buildings are domestic properties incorporating newly built complete houses and extensions to houses, which may be of any construction type.

Extensions can be connected to existing walls with timber wall plates or steel angles anchored to the walls.

Figure 1.3.1 Two Storey House Construction

Above 3 stories, for any building type, SIPs become less economical; as vertical and lateral loads accumulate down to ground level, often necessitating additional stiffening or strengthening structure. A common arrangement is to have SIP external walls and timber stud internal walls, sheathed if necessary, for strengthening or as a shear (racking) wall. This minimises the material cost. Where large internal space is desirable or internal load bearing walls do not line up with those above, beams comprising steel, timber or flitch beams can span between posts.

Figure 1.3.2 Interior Roof (*of* figure 1.3.1)

Figure 1.3.3 Completed House

Figure 1.3.1 shows a three storey house with a large footprint, clad in brickwork. **Figure 1.3.3** shows a one storey house with white render on render board; note the external breather and internal vapour control membranes in **figures 1.3.1** and **1.3.2** respectively.

1.3.2 Modular Units (Garden Offices, Accommodation)

Eminently suited to modular off-site construction - whether transported as complete units or partial assemblies erected on site - SIPs offer quality, great insulation and may be easily handled without the need for heavy mechanical lifting equipment. A 1.2mx2.4m typical OSB faced SIP panel is no more than 75kg in weight. The modular unit in **figure 1.3.5** weighs approximately 12 tonnes (12,000kg) fully clad.

Figure 1.3.5 shows the New Meaning Foundation workshop, where modules were built onto steel chassis. Walls are SIPs construction with joisted roofs with finished units 3.5x9.3m on plan with Kingspan 172mm TEK panels, supplied by Potton. The pods were fully fitted out as single bedroom homeless accommodation, then transported on flat bed trucks. 3.5m width units are permissible on UK roads with an escort. See **chapter 7**.

Figure 1.3.5 SIPs Units Off-site Assembly

Modular accommodation blocks in **figure 1.3.6** were fabricated from 'U' shaped floor and wall units, deriving their strength from joints formed from steel plates bolted to glulam wall and floor elements, which were connected to 'V' shaped roof assemblies on site. For each block three floor and roof segments were tied together with external grade ply nailed onto the glulam legs.

Figure 1.3.7 shows a three storey accommodation block concept which was designed with SIPs modules assembled solely with adhesive and screws, then site assembled with fixing brackets. It was demonstrated without any additional framing or lateral bracing. Platform access was to be provided by built on timber frames. The concept was envisaged as prefabricated units, complete with services (plumbing, electrical, heating and ventilation systems) prior to transportation and installation on site.

Figure 1.3.6 Modular Accommodation Block

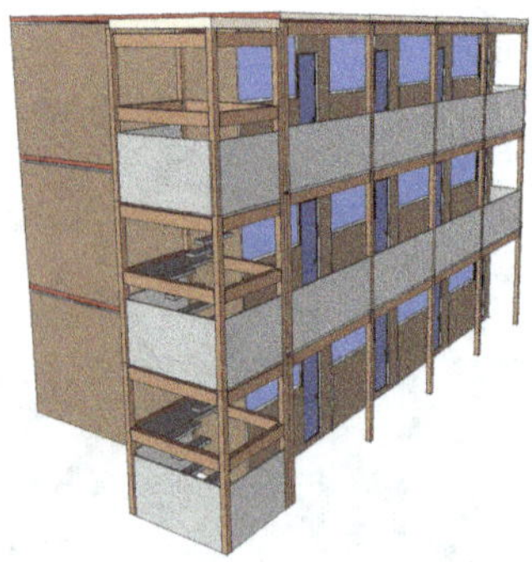

Figure 1.3.7 Multi-storey Scheme

1.3.3 Commercial, Education and Leisure

A growth in uptake of SIPs construction has thrived in the education sector, particularly in schools. With the growth of population and demand for classrooms and facilities for all ages, there have been many additions to existing infrastructure. The public sector is an enthusiastic supporter of sustainable buildings which are designed for minimal energy use. For larger spaces in 1 - 3 storey buildings, combined with glulam exposed beams, SIPs offers an attractive environment.

Figure 1.3.8 shows a school classrooms and assembly blocks in construction with glulam beams spanning onto SIPs perimeter walls.

Figure 1.3.9 shows a finished school building, timber clad, supported on a ground beam and screw pile system.

Figure 1.3.8 School Classroom in Construction

Figure 1.3.9 School Classroom Block

When designing schools and commercial buildings the issue of disproportionate collapse needs addressing. See the UKTFA Engineering Technical Bulletin No. 5, available from the Structural Timber Association (STA).

Figure 1.3.10 shows a two storey office showroom extension. It was designed for office floor loading with a large opening internally to give access to the existing workshops.
SIPs is suitable for stand alone commercial office space, and leisure use such as swimming pools.

Figure 1.3.10 Showroom - Office

1.3.4 Hotels and Apartment Blocks

A growing number of buildings with multiple residential unit and hotels have employed SIPs.

Figure 1.3.11 illustrates a built project for a charity that provides administration and apartments.
Hotels and apartment blocks that involve repetitive cellular rooms or self contained flats or studios may be constructed with SIPs party walls and joisted or panel floors (precast concrete panels or cross laminated timber panels). An example is shown in **figure 4.9.1**.

Figure 1.3.11 Residential Apartment Accommodation Complex

Alternatively, units may comprise stacked modules, assembled off-site, and lifted into their permanent location. Ground floor retail or amenities can be provided as conventionally constructed frames in steel or reinforced concrete. See **figure 1.3.7**.

1.3.5 SIPs as Infill Panels or Cladding Support

SIPs are commonly used in multi storey buildings, as insulated clad panels. An example of SIPs as clad panels in a challenging environment was for the replacement of a Nissen hut on an Antarctic camp at Port Lockroy, shown below in **figure 1.3.12.** The SIPs supplier and element designer was SIP Co Ltd.

Constructed by hand, SIP panels are fitted securely to arched steel frames, designed to withstand high wind and heavy snow cover. It is clad in corrugated steel panels. All structural elements had to be carried and installed by hand. See also in **figure 1.3.16**, Pelletron Building for another SIP panel infill example built into steel framing.

Figure 1.3.12 Port Lockroy Nissen Hut

1.3.6 Further Examples

To conclude this section, a number of successful projects highlight the versatility of SIPs construction. The writer was the structural engineering consultant in all of the examples which follow in this section. Each client employed a specialist SIP contractor and builder who were contracted to employ a 3D CAD designer and structural engineer as part of their package.

Figures 1.3.13 and **1.3.14** show a barn conversion .built by SIP Hus with 'before and after' photographs.

Figure 1.3.13 was taken from the rear. Steel framing was installed to brace the glazed opening and support the ridge beam.

Figure 1.3.15 shows a house designed and shipped from the UK and built by Bentley Projects in a stunning location in Switzerland.

The SIPs roof provides well insulated and sealed compartmentalisation, ideal in a challenging winter environment.

Figure 1.3.13 (top) & Figure 1.3.14 (above) Barn Conversion Before and After Completion

Figure 1.3.15 Alpine House Switzerland

The roof extends to form an overhang at one gable. This is built with timber joists sitting on a timber frame supplied and built by the client, where insulation is not necessary.

Careful planning for assembly and component size, and careful coordination with the client, was undertaken prior to erection.

Figure 1.3.16 Pelletron Building Cumbria

Figure 1.3.17 House in Cornwall

The futuristic building in **figure 1.3.16** is steel framed construction with SIPs panels supplied by SIP Co Ltd. infilling the façades and rainscreen panels fixed to the external face.

In **figure 1.3.17** a two storey gable for a house in Cornwall with large openings. To stabilise this elevation glulam frame posts and beams are inter-connected with large nailplate fixings, which provide sufficient moment (bending) capacity to achieve structural integrity. See detail **figure 8.3.18**.

A single storey school in Cornwall, shown in **figure 1.3.18** has large open plan spaces. Glulam purlins were arranged to split the SIP roof panel spans. Internal stud walls formed internal partitioning.

The roof overhang was supported at the edge by perimeter posts. The overhanging external roof is supported by timber joists rather than SIP, to save cost where insulation is not necessary.

Figure 1.3.18 School in Cornwall

Figure 1.3.19 House Interior - Open Plan Space

The open plan interior in **figure 1.3.19** is achieved by open plan glulam trusses supporting glulam purlins, which in turn provide the structural support for the SIP roof panels.

Rooflights are formed in panels by trimming with *timber splines* built into the sloping panels. See figure 1.4.3 for timber splines.

Below in **figures 1.3.20** and **1.3.21** are two hipped roof structures in different stages of construction built by SIP Hus. The hip joints are formed with different types of hangers. The hipped structure to the left has raking steel plated brackets fixed to the glulam ridge, which sits on the hip beams by means of slotted plates and stainless steel dowels. See **figure 8.3.9** in chapter 8.

To the right, the hip is formed with hidden fixings with interlocking 'male and female' concealed proprietary hangers. This type of connection is easier to construct than the connection opposite, which needs tight tolerances to achieve the connection. Hip rafters and ridge beams may be built into the SIP panels to reduce their spacial intrusion, noting that this compromises the insulation level of roofs.

Figure 1.3.20 Hipped Roof Underside

Figure 1.3.21 Hipped Roof Framing

1.4 Typical SIPs Assembly and Joints (as Detailed in Chapter 8)

The panels in **figure 1.4.1** is a typical arrangement made from 1.2m wide OSB strips. Based on a standard OSB panel, the length is 2.4m. Many UK manufacturers, such as Kinsgpan, Structural Building Systems or Hemsec, manufacture greater lengths up to circa 6.5m long; which has great benefits, particularly in roof construction.

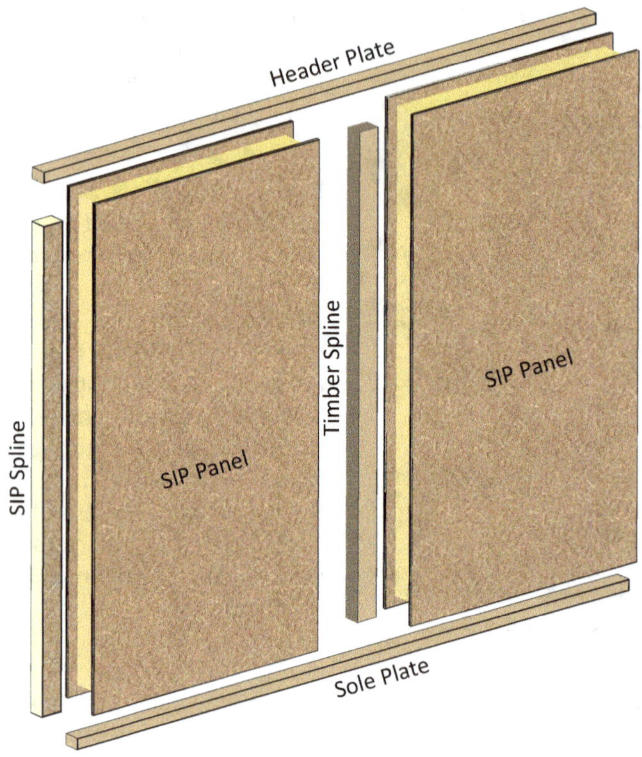

Figure 1.4.1 Typical SIP OSB Panel Assembly

Two wall panels are shown. At the base is a *sole plate* which is fixed to a base as illustrated in **figure 1.4.5** overleaf.

Wall panels are located on the sole plate, then nailed and bonded into place. Assembled panels are joined together at their edges with *spline joints*. Spline joints shown in **figures 1.4.2** and **1.4.3** are constructed with trimming timbers or shallower SIP strips, which are fitted to form interconnecting spline joints. Header plates or *head binders* trim the top. Recesses are formed during the manufacturing process around the full perimeter, typically 50mm deep, to allow trimming.

The spline joint in **figure 1.4.4** is commonly adopted in the UK, providing robust interconnection of panels.

Site assembly of spline joints is achieved by applying a bonding agent - adhesive or silicone - to the face of a spline, then nailing in place to compress the faces prior to the bonding agent setting. This seals the joint and also provides a strong shear interface, enabling assemblies of SIP panels to act as diaphragms: plate action giving wall or roof panel racking strength and rigidity. Each manufacturer has a preferred specification for sealants and nails.

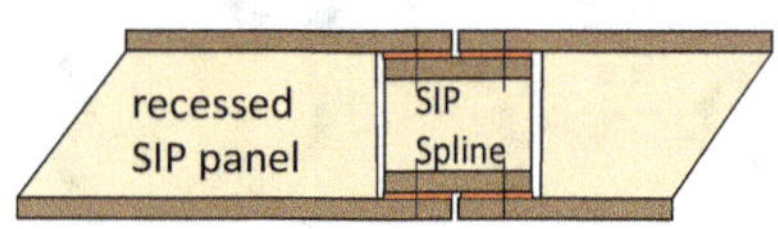

Figure 1.4.2 Typical SIP Spline Joint

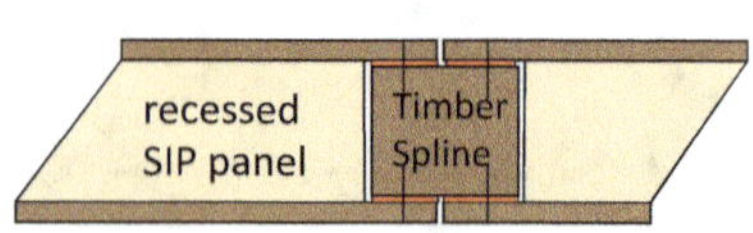

Figure 1.4.3 Typical Timber Spline Joint

At the base, walls are fixed down to the supporting substrate - ground slab, brick or block sleeper walls - onto soleplates. Figure 1.4.5 shows a simplified detail that shows a soleplate fixed through the DPM to the top of a supporting masonry sleeper wall with anchors at regular centres. Anchors may be expanding or resin anchors, or self tapping fixings designed for fixing to masonry or concrete, such as Tapcon or Thunderbolt. Anchors are necessary to hold the SIP structure in place. They resist any lateral (racking) actions, for example, wind forces, and, if necessary, any uplift loads that may result. Panels are lifted onto the sole plates, fixed into position, then nailed and bonded (as for a typical spline joint); additional straps may be fixed down to foundations or ground slab if needed. **Figure 6.2.4** illustrates a base detail with double soleplate.

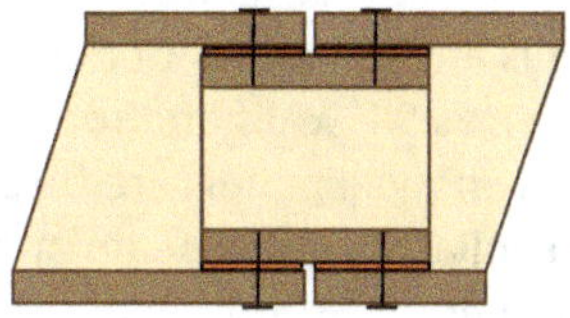

Figure 1.4.4 'As Built' Spline Joint

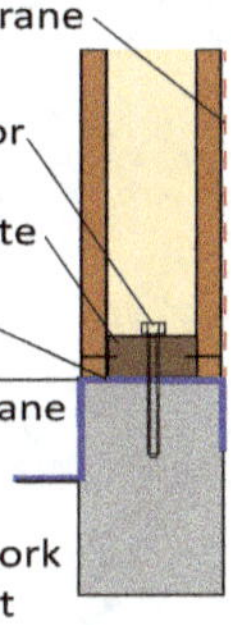

Figure 1.4.5 Sole Plate

Figure 1.4.6 shows a typical floor joint formed by overlapping the floor deck onto a wall header plate below. This ties in the floor diaphragm to the wall beneath and provides a construction deck for the builder to fix the upper floor sole plate into position by screws extending into the header below. The upper wall panels may now be installed.

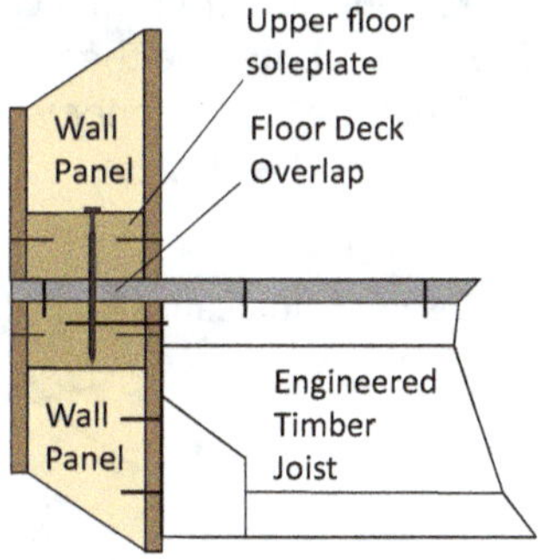

Figure 1.4.6 Floor Joint at Header

For additional bond, the upper floor soleplate may be adhesive bonded to the floor deck above prior to screwing. Engineered joists (timber flanges and lattice or OSB webs) are most commonly used in SIP construction with spans up to 6-7m.

Figure 1.4.7 illustrates a typical eaves joint is formed by fixing a profiled timber that is nailed and bonded to the header under off site, or on site if preferred. The roof panels are fixed to the wall header with *'SIP screws'*.

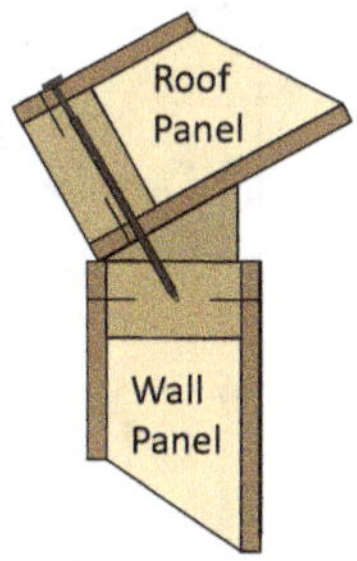

Figure 1.4.7 Eaves Joint

'SIP screws' - as they will be referred to in the book - are typically 6-8mm in diameter are specialist headed wood screws up to 350mm long. They are commonly fixed at 300 - 350mm centres (depending on design wind uplift conditions) in eaves, corners, wall to wall joints (shown in **figure 1.4.6**) and SIP to SIP connections. Examples of timber screws are the ESCR or SWW washer head screw, in the Simpson Strong-Tie catalogue, or Fastenmaster Headlok.

1.5 TYPICAL BUILT ASSEMBLY

Erection of SIPs buildings is essentially the same as timber frame construction. Figure 1.4.8 is an illustration of a typical building section. A one storey duo pitch frame comprises SIP walls and roof elements with a timber (generally glulam) ridge beam. SIP wall panels are fitted to a sole plate (see **figure 1.4.5**) on sleeper blockwork to a similar width. A beam and block with overlay of insulation and screed forms the ground floor. Ground bearing slabs and suspended timber floors are other options.

Two types of wall and roof cladding are shown: corrugated steel, timber shingles or render on board to the left. The right hand wall is faced with brickwork - with linking brick ties fixed to OSB faces - and the roof is clad with slate or clay tiles.

A full enclosure of a fully lapped breather membrane to the walls and roof panels is secured to the panels with nailed battens. Steel profile sheets or tiles may be fixed to the battens. A damp proof membrane (DPM) runs under the insulation layer and tucks under the bottom face of the sole plates at the finished floor level (FFL). The breather membrane then laps onto the DPM. A vapour control barrier may be fixed with battens to the inner wall faces if required. See **figure 1.3.2**. Internal faces are typically clad with skimmed plasterboard.

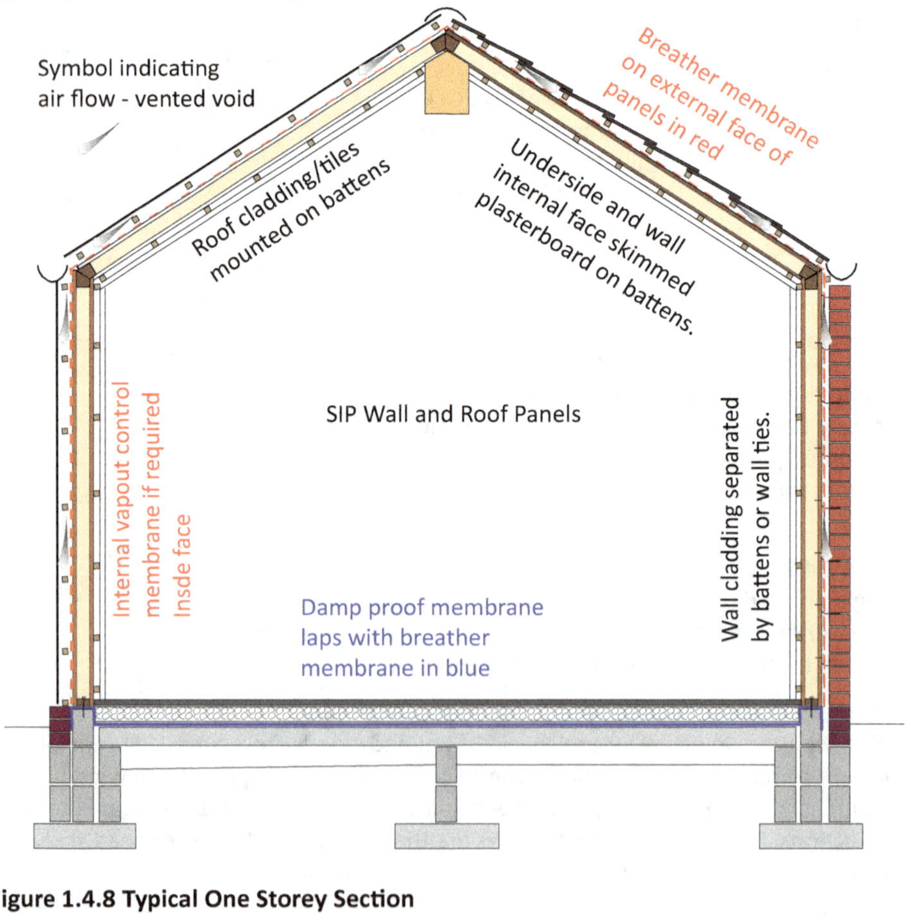

Figure 1.4.8 Typical One Storey Section

1.6 SIPS IN CONSTRUCTION

Whilst this book is intended as design guidance to structural engineers, it is worthwhile to summarise pertinent aspects of SIPs construction, which are essential for the engineer to appraise in the process of design. It also helps to assess risks so that health and safety or quality issues may be highlighted in the design documents or drawings.

1.6.1 Tie in Cross Walls

As a rule, all walls (masonry, SIP, timber stud or other) should be tied together at a junction to stabilise a building. Panel drawings should indicate connection details for connecting SIP panels together. Mostly long screws are used, and internal timber studs are located at joints to strengthen connections. See the plan view in **figure 1.4.9**.

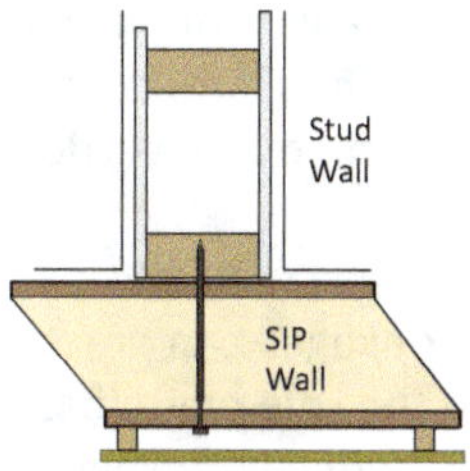

Plan View

Figure 1.4.9 Corner Junction

1.6.2 Tie in Roofs and Floor

Roofs and floors should be sufficiently tied together. SIP panels are fixed together with screws, which effectively tie structural elements together. Shown in **Figure 1.4.10** a typical internal wall joint is fixed through into the header plate of the wall below, which may be either SIP or stud construction, the upper wall sole plate.

The floor deck, such as chipboard, is sandwiched between the sole plate and header plate. **Figure 1.4.6** shows an external SIP wall fixed in a similar manner. This effectively ties floors and walls together.

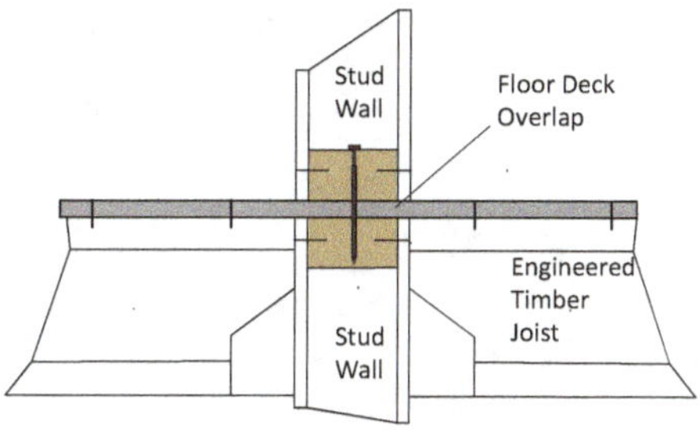

Figure 1.4.10 Floor Overlap at Internal Wall

1.6.3 Temporary Propping

Lateral wall propping should be considered as a temporary measure during construction at intervals of no less than 3.6m (3 panel width) as a rule with a timber sloping prop. The prop should be well secured to the timber header plate. See item 4 in **figure 1.4.14**.

1.6.4 Fire Resistance

SIPs and timber framed walls are generally faced with plasterboard on battens. Wall and floor junctions are sealed off when properly constructed. This forms a fire protected compartment internally, assuming that floors and ceilings are sufficiently fire protected. Roofs generally do not require special attention; for the most part under drawn with a fire resistant soffit. UK certified type F single skin plasterboard will provide 30 minutes of fire resistance. A double skin type F is sufficient for 60 minutes of fire resistance. These criteria ignore the fire resistance of SIPs panels.

1.6.5 Differential Movement/Shrinkage

Unlike timber frames, SIP panels, if kept dry, have very low shrinkage or expansion in use. During construction, panels should be weather protected if there is any risk of prolonged exposure to rain/dampness. Consideration should be given to differential shrinkage/expansion of any adjacent materials.

For example, brickwork and blockwork, if tied to an adjacent SIP wall, to a height greater than three stories, should have sliding ties to accommodate vertical movement.

In a wall with SIPs panels over a floor height shrinkage is, in practical purposes, limited to the sole plate (base timber) and header softwood shrinkage across the grain (circa up to 2-3mm maximum) - per storey. Typical shrinkage for stud walls in timber frame buildings is circa 6mm per storey.

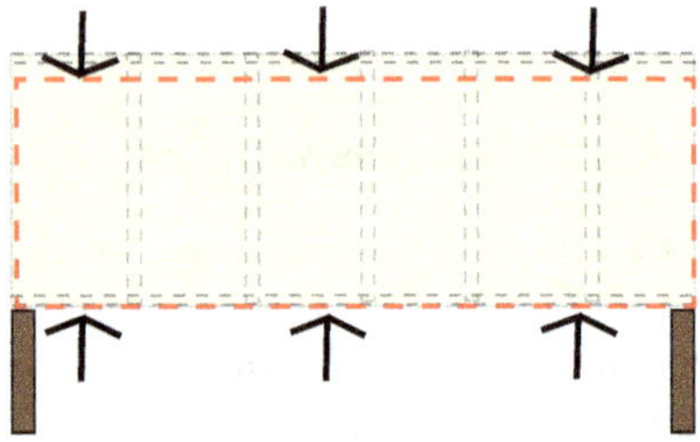
Shrinkage
(Long term drying out)

Figure 1.4.11 Shrinkage

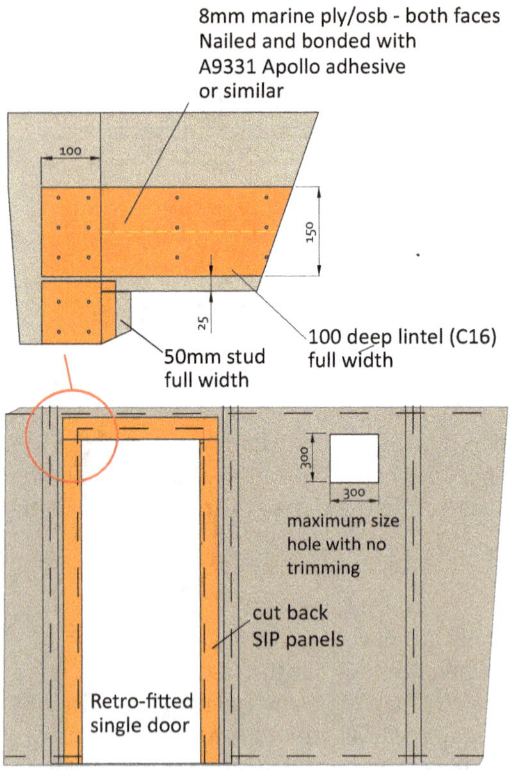

Figure 1.4.12 Site Cut Door Opening

1.6.6 Site Cut Holes

A manufactured panel with openings will generally have reinforcement in the form of built in timber joists or factory fixed angles.

In site cut openings, timber trimming joists or other forms of reinforcement, are fitted to trim around edges. If a hole is greater than circa 300mm square, the SIP insulation may be routed to form a recess and the timber inserted, adhesive bonded and nailed in place.

In the example in figure 1.4.12 a site cut opening is formed, and reinforced at the edges with OSB or plywood sheet that overlaps to form a robust remedial repair. Reinforcing strips of exterior grade plywood were nailed, and adhesive bonded providing an enhanced shear resistance by overlapping the lintel and adjacent panel.

1.6.7 Panel Joint Sealants

In any site finished joint, where OSB or Magnesium Oxide (MgO) face panels are connected to sole plates, slice joints are formed between panels and wall plates are fixed to the tops of walls. Generally, foam adhesive or silicone is applied to a face, and the OSB/MgO board is nailed or screwed to the face of the timber at regular intervals.

As a rule of thumb, a sufficient bond is indicated when adhesive seeps out of the edges of a joint.

The image shown in **figure 1.4.13** illustrates many aspects of SIPs construction on a house build: nailed and adhesive joints between panels, timber and steel lintels over openings, studs trimming and fixed into panel recesses to vertical edges (jambs) and upper wall located on a sole plate fixed through the floor deck into the wall below.

Figure 1.4.13 House SIPs First Floor Construction

1.6.8 Disproportionate (or Progressive) Collapse

Any part, or all, of a building may be susceptible to damage or catastrophic sequential collapse when not constructed in a robust way, for example, by vehicle collision or explosive blast.

It is incumbent on engineers to ensure that their design is safe in this respect. In the UK, the Structural Timber Association (STA) has published 'Engineering Bulletin 5' for timber frame structures, which addresses disproportionate collapse.

Building types (domestic, offices or schools, for example) are divided into categories (1, 2A and 2B). Reference is made to this in the book where relevant. In general, good practice construction, as outlined in sections 1.6.1 and 1.6.2, should avoid

disproportionate collapse in categories 1 and 2A. For category 2B, damage to primary load bearing elements, including walls, must be considered. This might entail analysis of blast loading or, alternatively, removal of load bearing walls. See **section 3.6.2**.

1.6.9 Roof SIP panels Building on Existing Masonry Buildings

A number of refurbishment projects involve fitting new SIP roof panels to existing masonry walls, which may include masonry gables. There are two important issues for the builder: potential lack of fit on gable ends and the possibility of pushing outwards and cracking eaves walls and gables when SIP panels are lifted into place. The roof should be properly fixed to restrain the gable in the long term. Temporary propping should be provided to prevent the latter.

Timber plates should be fitted and carefully checked for level and geometry. When roof panels are lifted and screwed to the sloping plates at the gables and eaves plates, the roof fitter should ensure a tight fit. Plastic shims as packing can be used where the existing walls are not perfectly level.

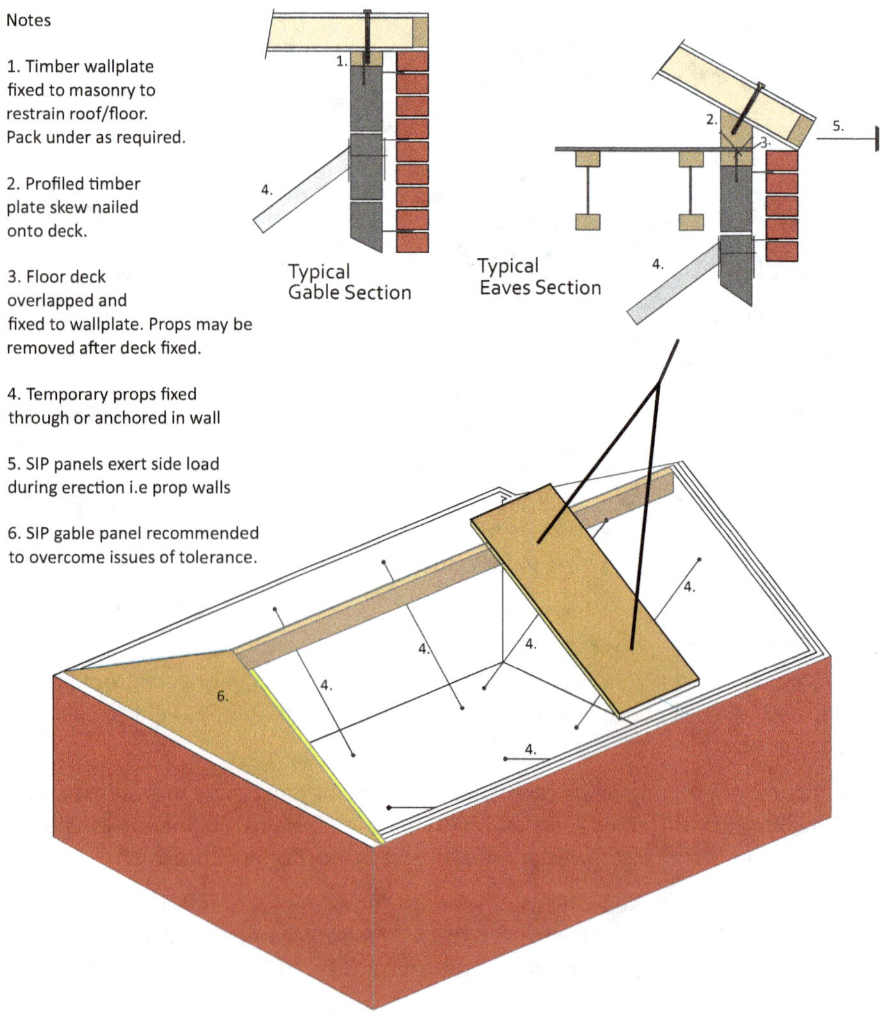

Figure 1.4.14 New SIP Roof Fitted to Existing Masonry Walls

1.7 RESOURCES AND REFERENCES

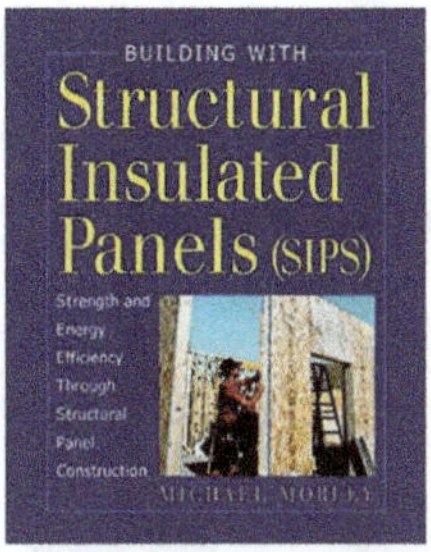

Morley M. (2000) Building with Structurally Insulated Panels (SIPs), The Taunton Press, 63 South Main Street, PO Box 5506, Newton, CA.

A comprehensive guide to practical construction, with references to properties, from USA in imperial units. Essential for those wanting a better understanding of SIPs construction.

Lstiburek J. (2008) Builder's Guide to Structurally Insulated Panels, SIPA, P.O. Box 39848, Fort Lauderdale, FL 33339

A thorough technical guide, by SIPA, which is packed with useful technical information including thermal, acoustic, condensation and building details.

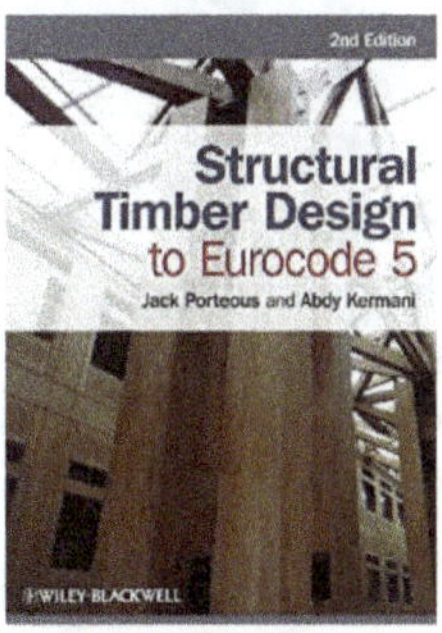

Porteous J. & Kermani A. (2013) Structural Timber Design to Eurocode 5, 2nd Edition, Wiley-Blackwell, John Wiley and Sons Ltd., 111 River St, Hoboken, New Jersey

Comprehensive text book for timber design to EC5 including reference to SIP racking, by Jack Porteus and Abdy Kermani of Edinburgh Napier University.

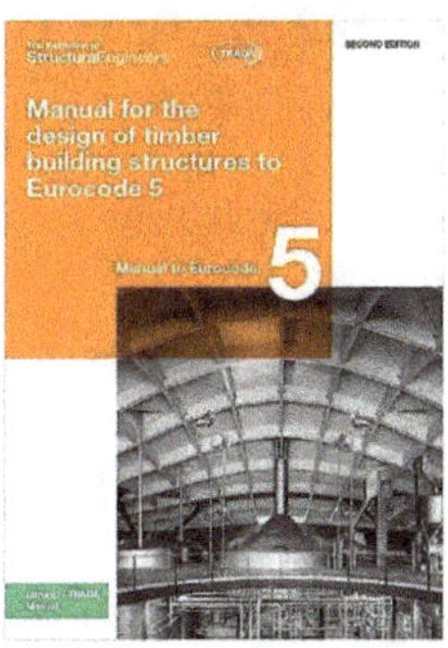

Trada/IstructE (2020) Manual for the design of timber building structures to Eurocode 5 (Second edition), 2nd Edition, The Institution of Structural Engineers, 47-58 Bastwick Street, London EC1V 3PS.

Useful reference for timber design to EC5.

Chapter 1

Other Information and Basis of Design

Nomenclature and units in the book

Dead loads are used to describe the self weight and service loading combined. Live Loads are used to describe the applied loading of the occupancy.

All units are SI units:

kN – forces
kNm – moments and torsion
m – spans and overall dimensions
kN/m (= **N/mm**) - line loads and shear flow/rolling shear
N/mm² - internal stress
mm – deflections

Limit State Design to Eurocode 5

Throughout, with the exception of racking, limit state design is followed, with partial load factors applied, and material ultimate capacities obtained from characteristic values. In the case of racking permissible racking loads are obtained from the relative BBA certificate for the panel being checked.

Factors can be applied so as to derive equivalent permissible (or unfactored) resistances, if working to permissible stress codes such as BS5268; albeit that this was officially withdrawn in 2010.

With deflection checks, modification factors (kdef) are applied. Since the insulation core contributes to the stiffness, instantaneous, medium term, long term and permanent values have been applied as the best that can be obtained industry standard. See **table 2.2.1.** This is not ideal, but observation for tests and real life structures has demonstrated that the values used are appropriate. Standardised testing and verification of shear modulus and creep movement for each core type should be the norm in ideal circumstances. This is discussed in **chapter 9**.

Other Information (available online)

SBS Design Guide (download from http://www.sipbuildingsystems.co.uk)
Structural guide, with load span tables and calculation method for bending, short to long term deflection, based on classical composite sandwich structure formulae.

"Engineered Design of Structural Insulated Panels", by Eric Tompos for the National Timber Association (pdhonline.com: *https://tinyurl.com/2f2vnuxa)*
Useful comprehensive set of structural formulae, based on imperial units:

Structural Timber Association (STA) (https://www.structuraltimber.co.uk/) Timber Development UK (https://www.timberdevelopment.uk)

SIPs and structural information with respect to timber and fixings, from these UK organisations.

Chapter 2

Bending (Flexure)

Bending or flexure of OSB SIPs panels from first principles using classical composite section formulae. SIP roofs - long term creep. Two way spanning. Pitched roofs. Finite element models. SIPs floors; short term and long term assessment. Allowance for panel voids. Continuous spans. Shear. Local core compression; line loads, point loads and screw uplift. MgO (magnesium oxide), steel and aluminium faced SIPs panels.

2.1 BENDING (FLEXURE)

SIPs panels are most commonly employed in pitched and flat roofs in domestic and commercial construction as a cold roof system. SIPs splines are the most thermally efficient, but longer spans are possible with timber splines. SIPs floors are used often in modular 'garden office' type buildings for floors that benefit from their insulative properties.

Roofs generally have lower live loads from snow or occasional access than floors. Of most concern is the phenomenon of long term sag, in which the 'creep' effect causes panels to sag progressively over a long period of time, until this stabilises after a period of years. Green roofs with a natural planted cover need careful consideration in design where dead loading is high.

Floor design must take into account the considerations of short term loading, serviceability checks for 'liveliness' or resonance and long term sag under dead loads. For both floor and roof design is governed mostly by deflection and not strength.

Other considerations for floors and roofs are the benefits of continuous panels, up to circa 6.5m long, and analysis of panels with significant voids.

Flexural analysis is undertaken in this chapter with the use of classical stressed skin sandwich panel principles concerning the outer faces and internal foam shear interaction.

Panels may be significantly stiffened by incorporating stiff timber, or other splines. This is a benefit structurally, but the designer must take into account the 'cold bridging' influence of the splines, which has the effect of increasing U values, which raises heat conductivity.

Figure 2.1.1 Panel Bending Under UDL

2.1.1 Analytical Method of Panel Design in Bending

Bending Strength

The method used for analysing the flexural action based on the principles of sandwich panels. As the faces are fully bonded to the foam - whether OSB, steel, aluminium or MgO - their mode of resistance to flexural action is primarily axial: tension and compression on the opposite face. See **figure 2.1.2**.

Deflection (or Deformation)

Sandwich panel interaction comprises two components: the elastic deformation of the faces in tension and compression, which is at a maximum in midspan for a uniformly distributed load, and the shear deformation of the foam core where shear is at a maximum (at the support for a UDL). See **figure 2.1.3**.

Shear Strength at Support

Shear strength for panels with sip splines (referred to as unreinforced) is related to the shear strength combined for the faces and foam. When a panel has timber splines (referred to as reinforced) the shear is enhanced by the shear strength of the timber splines. In general use for roof panels and floor panels, shear failure is not a critical consideration, but should be considered if particular loading conditions, such as high local point loads.

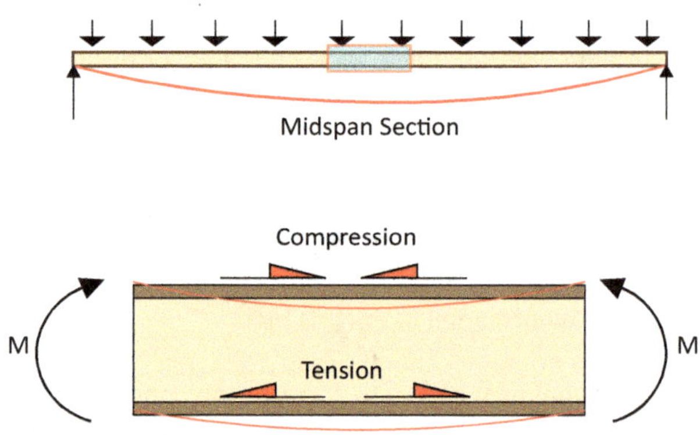

Figure 2.1.2 Panel Stressed Skin Action Midspan

Local Point Loads

Local point loads or 'pull through' modes of failure under screw or bolt heads under uplift load conditions are limited by the compression strength of the foam and the thickness of the face.

Also of consideration is the issue of brittle failure for MgO boards and local buckling failure of thin steel or aluminium faces. Strength in relation to OSB faces will be considered in **section 2.2**.

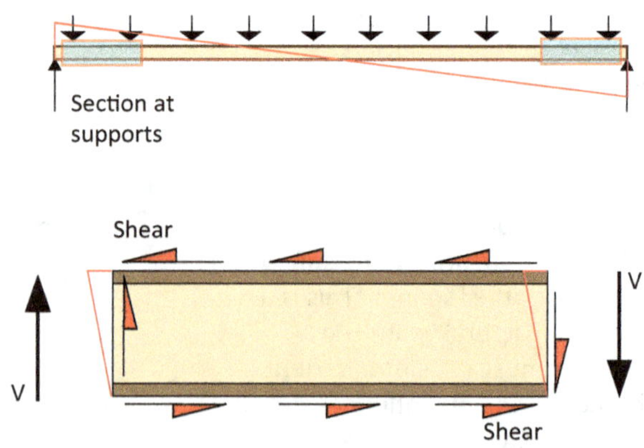

Figure 2.1.3 Panel Foam Shear at Supports

2.2 BENDING (FLEXURAL) STRENGTH CHECKS

For the purposes of checking SIPs panel for bending strength alone, the foam core is assumed to have an insignificant effect. For OSB faced panels with SIP splines (reference **figure 1.4.2**, chapter 1) the outer faces are the sole contributor to bending strength. When OSB panels have timber splines - referred to as reinforced - the panel is treated as a composite *stressed skin* section with timber web and OSB flanges, which are limited in width by the shear lag effect.

2.2.1 Bending Strength of Panel with SIP Splines (Unreinforced)

In the panel section in **figure 2.2.1** a UDL acts on a panel with span, L, with uniformly distributed load (UDL), w, and width, Bp. From the following parameters the bending resistance, to Eurocode 5, of the skins may be calculated:

Thickness of faces = t
Equal thickness faces will be assumed throughout the book.
Depth of panel = d_p, and Depth of Core = d_c
Effective depth, $d_{eff} = d_p - t = (d_p + d_c)/2$
Characteristic resistance in tension of face, $f_{t,0,k}$
Characteristic resistance in compression of face, $f_{c,0,k}$
Noting that the OSB sheet is orientated with outer grain parallel to the span

Partial factor for material property, Υ_m, where $\Upsilon_{m\,(OSB)} = 1.2$
Modification factor for load duration, K_{mod} (assumed service class 2)
(K_{mod} values: permanent = 0.30, long term = 0.40, medium term = 0.55 and instantaneous = 0.90)

For Kmod, service class 1 and 3 (external) refer to Eurocode 5, Table 3.1.

Design bending resistance in tension, $M_{d,t} = \dfrac{f_{t,0,k}\,.\,B_p.\,t.\,K_{mod}.\,d_{eff}}{\Upsilon_m}$

For $M_{d,c}$ substitute $f_{c,o,k}$ ($f_{t,0,k} < f_{c,0,k}$ generally. Therefore $M_{d,t}$ is critical)

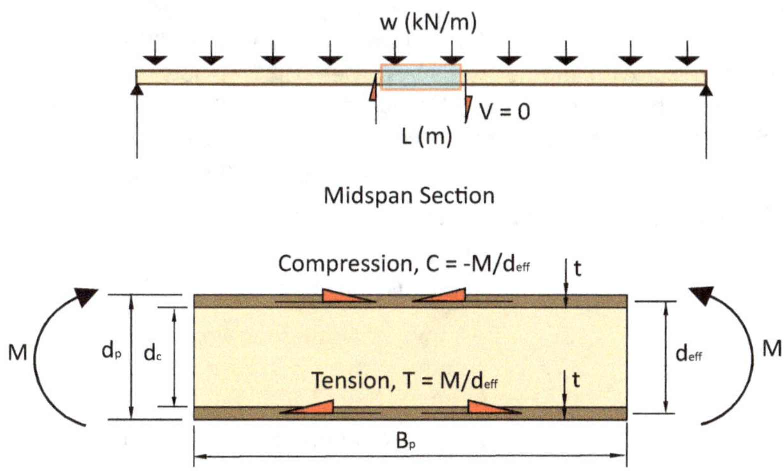

Figure 2.2.1 Panel Section at Maximum UDL Moment Position (Centre)

2.3 STIFFNESS OF PANEL: FROM BENDING AND SHEAR COMPONENTS

Stiffness is influenced by the bending component - maximum tension/compression in skins - and shear component of the core, which is a maximum at the supports for a UDL condition. For long and medium term loads stiffness is influenced by 'creep' in the material: the longer the loading, the greater the creep effect. Most often, the critical design consideration for roof panels with relatively light dead loads, is long term creep. For floor panels, with relatively high live loads, the critical design condition can be medium term ultimate conditions. Another important phenomenon for floors is the 'liveliness' or resonant response to instantaneous live loading[1]. In other words the natural frequency could be a governing factor.

For a panel with UDL, w, and span L, the deflection midspan may be expressed as:

$$\text{Deflection, } d_{max} = \frac{5.w.L^4}{384.E_f.I_f}\,.\,(1 + k_{def,f}) + \frac{w.L^2}{8.A_c.G_{c(B)}.}\,.\,(1 + k_{def,c})$$

To the left is the *bending component*, and to the right the s*hear component*.

E_f = Young's Modulus of skins (faces) - typically 3800 N/mm² (Based on $E_{t,0,mean}$ - EC5) *assuming compression modulus = tension modulus*
This should be confirmed by referring to the manufacturer's data.

I_f = Moment of inertia of faces ($I_f = (d_p^3 - d_c^3).B_p/12$)
A_c = Area of foam core (*alternatively* substitute $A_{eff} = d_{eff}.B_p$ from Allen[2])
$G_{c(B)}$ = 'Flexural' shear modulus of foam core derived from shear modulus, G_c
$K_{def,f}$ = Load duration factor for the face
$K_{def,c}$ = Load duration factor for the core

Typical K_{def} - Load Duration Factors (Service Class 2)

Table 2.2.1 summarises typical values for long term Kdef. They were obtained from different sources. For timber (including glulam, LVL and Kerto) and OSB/3 long term values were taken from Eurocode 5 table 3.2, assuming service class 2. Foam creep coefficients - assumed conservatively to be equivalent to Kdef values - were quoted by Eric Tompos (Vice President, NTA, USA) from a paper presented in the Journal of Structural Engineering (1997)[3]: 7.0 for PUR foam and 4.0 for EPS/XPS polystyrene foams. These concur with another study by Davies (1987)[4] (a lower coefficient of 3.0 is quoted for polystyrene cores, but 4.0 is indicated). TR019[5] E.3.2.3 quotes Kdef for all types of core to be 7.0, which was adopted for PUR foam but not for EPS/XPS; as this is regarded to be unreasonably conservative, in the light of Tompos et al and Davies.

Typical $G_{c(B)}$ - Flexural Shear Modulus Values for the Core

The basic core shear modulus, G_c, values are dependent on the core density and type. *Depending on the SIP panel manufacturer,* G_c may be assumed as follows:

PUR foam core:	G_c = 1.7 N/mm² (3.5N/mm² for Kingspan SIPs)
EPS foam core:	G_c = 2.5 N/mm²
XPS foam core:	G_c = 4 N/mm²

From EOTA, TR019[5], C.2.1.2, ignoring OSB contribution, $G_{c(B)}$ may be calculated:

$G_{c(B)} = G_c.\dfrac{d_{eff}^2}{d_c^2}$ ...for *simplicity* $G_{c(B)} \approx 2.3$ throughout for PUR foam from Rungthonkit[6]

Material	Instantaneous	Short Term 200hours (8 days)	Medium Term 2000 hours (84 days)	Long Term 100k+ hrs (10+ years)	Permanent 220k+ hrs (25+ years)
Timber, Glulam, LVL	0	0.2	0.4	0.6	0.8
OSB/3 ($K_{def,f}$)	0	0.5	1.1	1.6	2.3
PUR foam ($K_{def,c}$)	0	0.7	2.4	5.0	7.0
EPS/XPS foam	0	0.7	1.2	3.0	4.0*

Table 2.2.1 Typical K_{def} Load Duration Factors

**note: EPS/XPS factors from Tompos[8]*

2.3.1 Comments with Respect to Stiffness Calculations

There are many variables and limited long term test data for bending and creep. The parameters presented will cover the most available SIP panels in the absence of specific data. Some relevant comments are summarised below:

- As a rule, the following generally applies: dmax < L / 250 (from Table NA1 BS EN 1990) and L < 13-15mm, whichever is the lesser value. Long term deflection is frequently more critical. If short term deflection exceeds this criterion, part or all of the live load might be omitted and checked again.
- There is potential for reduction in K_{def} factors in many roofs and floors as shown by long duration creep tests at Eindhoven University[15]. See **section 9.2**.
- Floor deflection under instantaneous dead load should not exceed circa 6mm[6] as a guide. This is to avoid excessive liveliness or resonance. See **section 2.3.2.1**.
- Service class 2, from Eurocode 5, is defined by a moisture content not exceeding 20% and a relative humidity greater than 85% for a few weeks a year. Refer to Eurocode 5 for k_{def} for service classes 1 and 3. *A damp OSB face can cause significantly increased creep*; breather membranes and ventilation of voids under roof finishes are very important consideration.
- The assessments assume one way spanning of panels. When the parallel to perpendicular span ratio is in the region 1-1.5 two way spanning action can stiffen panels significantly. See **section 2.8** for an assessment of 2 way spans.
- Panels which are part of pitched roof installations, in a similar manner to joisted roofs with partial to full horizontal restraint, may have reduced bending owing to partial axial compression within the panels. Finite element models (FEM) are useful for verification of pitched roofs, as discussed in **section 2.8.1**.
- A more comprehensive European document for assessing stressed skin wood based panels, including 'load bearing foam' was published, known as TR019[5]. The simplified formulae presented herein for unreinforced and reinforced panels are sufficient for general use, and have been verified against test data available in the Rungthonkit paper[6] and from feedback from built projects.
- For panels with multiple supports, continuity can be taken into account by overlapping panels for reduced deflections. This is particularly useful for panels on roofs with intermediate purlins and on flat roofs with intermediate supports; discussed in **section 2.9**.

2.3.2 Deflection Checks: Load Combinations (Dead and Live)

In general use there are three load combinations which are commonly undertaken for panel deflection checks:

- Short term (dead and full live load combined.
- Medium term (dead load and full or *partial* live load).
- Permanent (dead load alone).

SIP floor analysis will generally involve consideration of all three cases. SIP roof analysis will be mostly concerned with the third case - permanent creep under dead load.

2.3.2.1 Short Term Deflection Checks

Since this formula is largely concerned with the performance of floors, if incorporated into a spreadsheet or mathCAD type formulation, it can readily be included for comparison with longer term analyses.

From **table 2.2.1** values for $K_{def} = 0$ for foam and face skin.
G_k (dead load) is combined with full Q_k (live load):

From **section 2.3** formula, short term Deflection, $d_{max} = d_{max}(G_k) + d_{max}(Q_k)$
applying $k_{def} = 0$ *for both core and face.*

'Liveliness' or Resonance Check

A reasonable estimate of fundamental natural frequency, for a UDL on a simply supported panel, may be found by applying the formula:

F_n (fundamental) = $C/\sqrt{dfn}$ ...where the constant, C = 17.7

$dfn = d_{max}$ assuming dead load combined with 10% live load (Gk + O.10QK).

Target F_n > 7 - 8Hz (for typical damping, circa 2%). See reference[1].

2.3.2.2 Medium Term Deflection Checks

This is generally a concern for roofs with snow loading. A 'quasi-permanent factor (ψ_2) for variable action' may be applied to the live load to take into account its limited duration.

From **section 2.3**, medium term Deflection, $d_{max} = d_{max}(G_k) + \psi_2 . d_{max}(Q_k)$
*applying K_{def} medium term from **table 2.2.1** for both core and face.*

$\psi_2 = 0.20$ for snow loading on roofs
$\psi_2 = 0.30$ for live loading on floors

This is considered reasonable when considering that the K_{def} values in **table 2.2.1** are quoted for a duration of 2000 hours (2-3 months). Typical live loading on office and domestic floors are in the region 0.50 kN/m² from a report published for Stanhope[7] in which actual office loadings were investigated.

2.3.2.3 Permanent Deflection Checks

Live loads are ignored for the purposes of undertaking permanent creep calculations under self weight. For roofs, this criterion most frequently governs the design.

From **section 2.3**, permanent *panel* deflection, $d_{max} = d_{max}(G_k)$
*applying permanent k_{def} from **table 2.2.1** for both core and face.*

2.3.4 Sloping Panels

When evaluating pitched roof panels, the slope length of a panel is applicable. The load component, ws, perpendicular to the face is substituted for w.

The length of panel in the formula given in **section 2.3** should be substituted for the slope length, Ls, in **figure 2.3.1**:

$L_s = L/\cos\theta$
$W_s = L.\cos\theta$

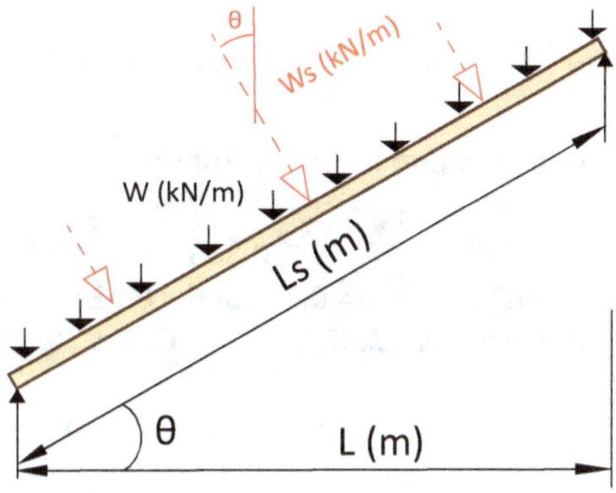

Figure 2.3.1 Slope Length of Panel

2.3.5 Example Panel Calculations (SIP splines/Unreinforced)

142mm Pitched Roof Panel Check

A 142mm deep SIP panel with 15mm thick OSB faces has a 30° slope, spanning horizontally 2.5m with dead load (G_k) and live (snow) load (Q_k) as shown in **figure 2.3.2**. The PUR foam core sits between OSB/3 faces:

$f_{t,0,k} = 7.7\text{N/mm}^2$
(compression strength is higher)

Young's modulus, E = 3800 N/mm²

Flexural shear core modulus,
$G_{c(B)} = 2.3\ \text{N/mm}^2$

Bending strength, followed by instantaneous, medium and long term deflections, will be checked based on the parameters hitherto presented in this section.

Calculations will be based on 1m width of panel for the UDL loading.

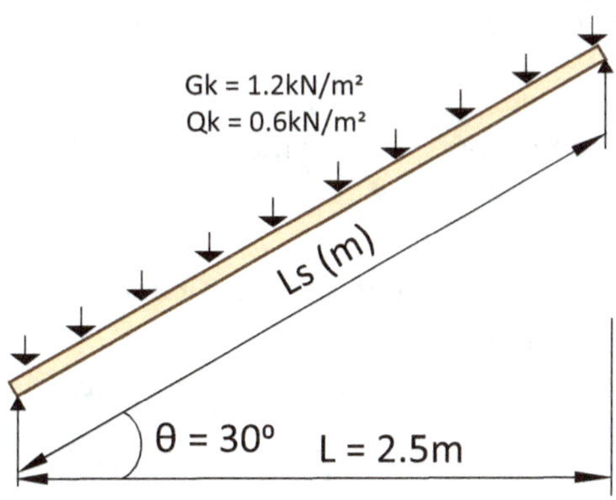

Figure 2.3.2 142mm Pitched Roof Panel

Design Bending Resistance in Tension

$$M_{d,t} = \frac{f_{t,o,k} . B_p . t . K_{mod} . d_{eff}}{\gamma_m}$$ .... from **section 2.2.1**.

Assuming medium term conditions for snow with no reduction:
K_{mod} (OSB/3, service class 2) = 0.70
$d_{eff} = d_p - t = 142 - 15 = 127$mm

$M_{d,t} = 7.7 \times 1000 \times 15 \times 0.70 \times 127 / 1.2 / 1\times10^6 = 6.7$kNm/m (1m width)

Ultimate Bending Moment Applied

Slope length, Ls = L/cos30° = 2.5/0.866 = 2.9m,
Loading on slope, $G_{ks} = 1.2\times0.866 = 1.04$ kN/m
Loading on slope, $Q_{ks} = 0.6\times0.866 = 0.52$ kN/m
Ultimate load, $W_u = 1.35.G_k + 1.50.Q_k = 1.35\times1.04 + 1.50\times0.52 = 2.2$kN/m

$M_u = W_u.Ls^2/8 = 2.2 \times 2.9^2 / 8 = 2.3$ kNm/m << 6.7

Design bending resistance is significantly greater than applied ultimate moment.

Instantaneous Deflection from Section 2.3

Deflection, $d_{max} = \frac{5.w.L^4}{384.E_f.I_f} . (1 + k_{def,f}) + \frac{w.L^2}{8.A_c.G_{c(B)}} . (1 + k_{def,c})$

($d_{max} = d_{max(Gk)B} + d_{max(Gk)V}$ - *sum of bending and shear components*).

Calculate moment of inertia of faces, per 1m width, $I_f = (dp^3 - dc^3).Bp/12$

$d_c = d_p - (t1 + t2) = 142 - 2\times15 = 112$mm.....therefore

$I_f = (142^3 - 112^3).1000/12 = 121\times10^6$ mm^4

Area of core per 1m width, $A_c = B_p.d_c = 1000 \times 112 = 112\times10^3$ mm^2

From **table 2.2.1**, $K_{def} = 0$ for core and faces. Calculating deflection with dead and live components separately simplifies calculations for medium and long term assessments.

Instantaneous Dead Load Deflection

Dead load per 1m width, $G_{ks} = 1.04$ kN/m^2 x 1m = 1.04kN/m
Ls = 2.9m (substitute for L in equation above), E = 3800 N/mm^2

Bending Component (Left Hand Part of Equation)

Maximum deflection, $d_{max(Gk)B} = 5\times1.04\times2900^4/(384\times3800\times121\times10^6)\times1/(1+0)$
$d_{max(Gk)B} = 2.1$mm

Shear Component (Right Hand Part of Equation)

PUR foam core: $G_{c(B)}$ = 2.3 N/mm² from **section 2.3**.

$d_{max(Gk)V}$ = 1.04x2900²/(8x112x10³x2.3) = 4.2mm

Instantaneous Live Load Deflection

Live load per 1m width, Q_{ks} = 0.52 kN/m² x 1m = 0.52kN/m

Since live load is half the value of dead load, then the deflection is 50% of dead load deflection, therefore:

$d_{max(Qk)B}$ = 2.1mm/2 = 1.0mm
$d_{max(Qk)V}$ = 4.2mm/2 = 2.1mm

Instantaneous Dead and Live Deflection Combined

Total instantaneous deflection for both dead and live is from the superposition of the four components:

$d_{max(Inst)} = d_{max(Gk)B} + d_{max(Gk)V} + d_{max(Qk)B} + d_{max(Qk)V}$

$d_{max(Inst)}$ = 1.9 + 4.2 + 1.0 + 2.1 = 9.2mm (Ls/315 > L/250)

Medium Term Dead and Live Deflections

Having calculated instantaneous deflections, these may be pro-rata calculated using appropriate K_{def} factors.

Medium Term Dead Load Deflections (K_{def} medium term)

From **table 2.2.1**, $K_{def,f}$ (medium term) = 1.1, $K_{def,c}$ (PUR/medium term) = 2.4

$d_{max(Gk)B.med} + d_{max(Gk)V.med}$ = 2.1x(1+1.1) + 4.2x(1+2.4) = 4.4 + 14.3 = 18.7mm

Medium Term Live Load Deflections (K_{def} short term)

Since live loading is based on a snow load, in the UK, for instance, it is reasonable to assume it is a short term condition. Therefore, short term deflection factors may be applied.

From **table 2.2.1**, $K_{def,f}$ (short term) = 0.5, $K_{def,c}$ (PUR/short term) = 0.7
$d_{max(Gk)B.med} + d_{max(Gk)V.med}$ = 1.0x (1 + 0.5) +2.1 x (1+0.7) = 1.5 + 3.6 = 5.1mm

Medium Term Dead and Live Deflection Combined

$d_{max(Med)} = d_{max(Gk)B} + d_{max(Gk)V} + d_{max(Qk)B} + d_{max(Qk)V}$
$d_{max(Med)}$ = 18.0 + 5.1 = 23.1mm (Ls/124) ...*high*

Permanent Deflection Dead Load Only (K_{def} long term)

From **table 2.2.1**, $K_{def,f}$ (permanent) = 2.3, $K_{def,c}$ (PUR/permanent) = 7

As noted in **section 2.3.2.3** live loads may be ignored.

$d_{max(Long)} = d_{max(Gk)B} + d_{max(Gk)V}$
$d_{max(Gk)B.long} + d_{max(Gk)V.long}$ = 2.1x(1+2.3) + 4.2x(1+7) = 6.9 + 33.6 = 40.5mm
span/depth ratio = 2900 / 40.5 = 72, *expressed as L/72 high*

Summary of 2.5m Span SIP Panel Deflections

A summary of calculated deflections from previous pages is shown in **table 2.3.1**

Material	Instantaneous	Medium Term 2000 hours (84 days)	Permanent 220k+ hrs (25+ years)
Dead $d_{max(Gk)B}$	2.1mm	4.4mm	6.9mm
Dead $d_{max(Gk)V}$	4.2mm	14.3mm	33.6mm
Live $d_{max(Qk)B}$	1.0mm	1.5mm	0
Live $d_{max(Qk)V}$	2.1mm	3.6mm	0
Total Deflection	9.4mm	23.8mm	40.5mm

Table 2.3.1 Summary of Deflections (142mm panel, 2.5m span, 30° pitch)

From the moment resistance calculation and **table 2.3.1** in summary:

- Strength is not an issue. The resistance $M_{d,t}$ = 6.7kNm/m, is greater than the applied design moment of 2.3kNm/m.
- In the short term, the deflection is circa 25% of the permanent sag. The critical condition is long term sag.
- The creep coefficient and shear modulus of the foam core have a significant influence on overall stiffness.
- Calculations were carried out for different spans at this pitch. A 1.5m span gave medium term deflection 7.5mm and permanent deflection of 13.5mm (Ls/215). It is possible that 13mm could be accepted if it is considered that water would not pond at this pitch, and visually this was acceptable.
- Panel stiffness could be significantly increased by installing timber splines in lieu of SIP splines to reinforce the panels. See **section 2.4**.
- If the panel comprised two equal 2.5m spans with a solid central support the permanent deflection would be 40% of the simply supported deflection at 16mm, indicating the benefit of continuity.
- If a continuous panel of two equal 2.5m spans was supported centrally by a purlin, the purlin deflection would cause greater sag than the continuous condition with solid support. Purlin interaction must, therefore, be considered.
- 142mm Kingspan panels, assuming $G_{c(B)}$ = 4.5N/mm^2 yields reduced deflections.

175mm Floor Panel in a Garden Office Spanning 2.5m

SIP floors are rarely used in domestic dwellings except in smaller garden offices. The calculation methods in the previous example were carried out for the garden office floor, as shown in **figure 2.3.3**. The floor comprises 175mm deep SIP panels, with PUR core and 11mm OSB/3 faces.

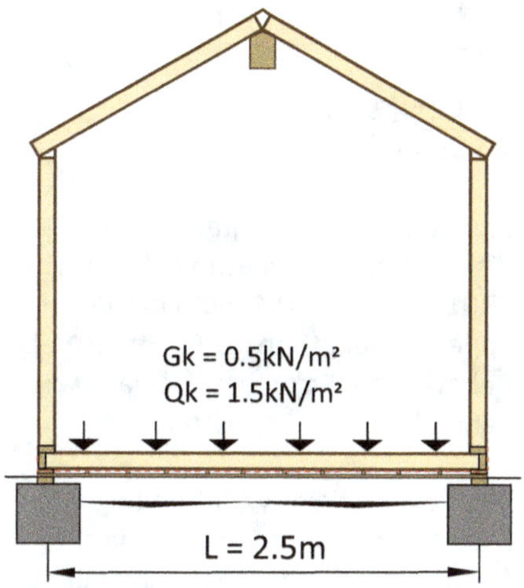

Figure 2.3.3 Garden Office Floor

The floor sits on a strip footing with sub-floor void. Each panel is underdrawn with an overlapping damp proof membrane, battened over with WBP waterproof coated plywood fixed to the underside.

It would be a wise precaution for the panels to have treated timber splines, which would carry the floor in the event of water impregnation of the OSB underside, noting that a marginally raised U value will result from cold bridging of the joists. OSB is prone to disintegrate in prolonged damp conditions. Following the same method for assessing deflections, the following deflections were calculated. *Note that no reduction in live load and class 2 was assumed.*

Material	Instantaneous	Medium Term 2000 hours (84 days)	Permanent 220k+ hrs (25+ years)
Dead $d_{max(Gk)B}$	0.5mm	1.0mm	1.5mm
Dead $d_{max(Gk)V}$	1.1mm	3.8mm	8.9mm
Live $d_{max(Qk)B}$	1.4mm	2.0mm	0
Live $d_{max(Qk)V}$	3.3mm	5.7mm	0
Total Deflection	7.5mm	12.4mm	10.4mm

Table 2.3.2 Summary of Deflections (175mm floor panel, 2.5m span, 0° pitch)

In summary, the instantaneous deflections are within acceptable limits. The medium term sag is equal to L/202, but if live has a quasi-permanent factor, $\psi_2 = 0.30$, then the total deflection would reduce to 7.1mm (L/352). The permanent sag (L/240) is greater than the acceptable criterion of L/250, but was accepted by the client with flexible floor finishes.

Fundamental natural frequency for the floor was calculated as follows:

Instantaneous deflection[5] (G_K + 10%Q_k) = 2.4mm, then $F_n = 17.7/\sqrt{2.4} = 11.4\text{Hz} > 8\text{Hz}$

Bending resistance at 6.4kNm/m was greater than the applied 2.3kNm/m.

The 175mm panel is satisfactory, assuming class 2 conditions are maintained.

2.3.6 Panels with Point Loads: Bending, Deflection and Shear Equations

Equivalent equations for point loads spread evenly across the panel conditions, compared with an applied UDL, as outlined in **section 2.3**, is shown below for a central point load, **figure 2.3.4**, and for two point loads at third span, in **figure 2.3.5**. Refer to **section 2.3** for parameters employed.

Beam with Central Point Load

$$\text{Deflection, } d_{max} = \frac{F.L^3}{48.E_f.I_f}.(1 + k_{def,f}) + \frac{F.L}{4.A_c.G_{c(B)}}.(1 + k_{def,c})$$

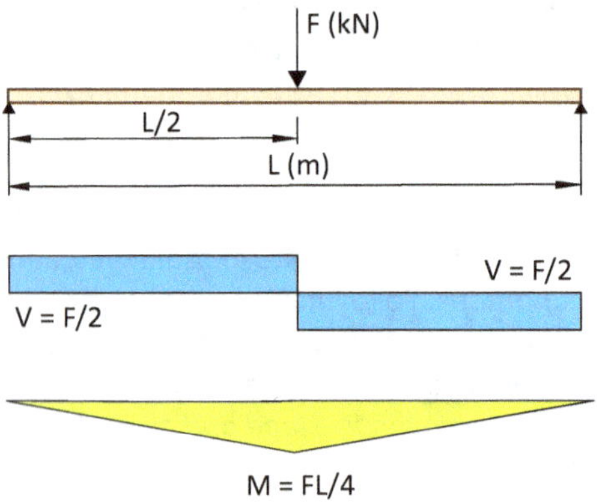

Figure 2.3.4 Panel with Central Point Load - Shear and Bending

Equations are presented for three different analysis methods (Allen[2], TR019[5] and BS EN 14509) in the paper by Rungthonkit[6] P, "Structural Behaviour of Structural Insulated Panels".

The flexural rigidity (or bending stiffness) of the OSB faces, and the foam core are ignored in the equation in **section 2.3**, above and below, on the basis that their influence is negligible.

Beam with 2 Point Loads at Third Span Position

$$\text{Deflection, } d_{max} = \frac{23F.L^3}{648.E_f.I_f}.(1 + k_{def,f}) + \frac{F.L}{3.A_c.G_{c(B)}}.(1 + k_{def,c})$$

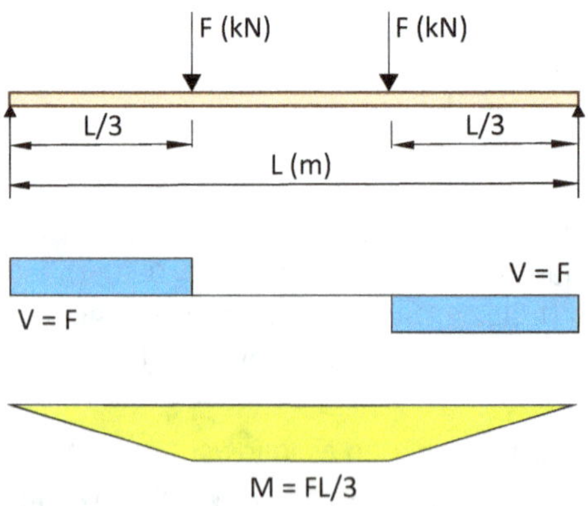

Figure 2.3.5 Panel with 2 Point Loads at Third Spans - Shear and Bending

....Alternatively, substitute A_{eff} for A_c, where $A_{eff} = B_p.d_{eff}$ from Allen[2] in both equations.

For analysis of roofs, an example of point load appraisal would be where roof plant, such as air handling units or fans sit on the roof panels, more commonly for commercial projects, such as offices or hotels. Point loads might occur, for instance, with heavy kitchen units or boilers. This could be checked for a single panel with a load assumed spread across the width. For local shear check, **section 2.12**.

2.4 Panels with Solid Timber Splines (Reinforced)

By the addition of timber splines, panel stiffnesses may be significantly increased. Calculations employ stressed skin panel equations where the panel section properties for instantaneous and longer term loading are primarily influenced by the composite actions of timber splines and panel faces. *The influence of foam creep is assumed to be negligible owing to its containment under the action of flexural loading.*

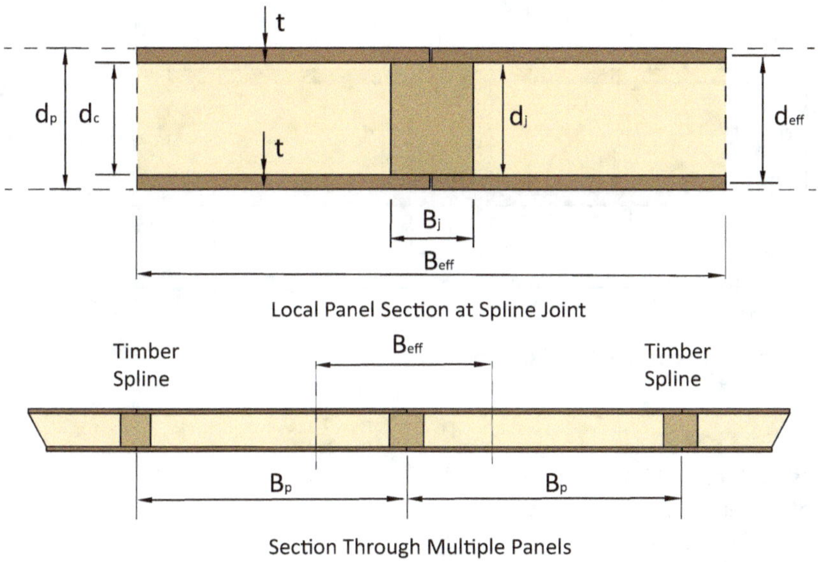

Figure 2.4.1 Panels with Solid Timber Splines (Reinforced)

Figure 2.4.1 illustrates multiple panels with solid timber splines. It is noteworthy that the splines may be engineered timber joists or steel hollow sections, which are commonly adopted as reinforcement. The principle of composite stressed skin panels in the equations that follow may be adapted for splines other than solid timber.

2.4.1 Effective Width of Panel, B_{eff} (Shear Lag Effect)

The phenomenon of shear lag limits the breadth of faces acting compositely with the reinforcing splines in a similar way to wide flanged beams. In a SIP panel bonding of the foam to the faces means that the effective breadth (or width) is wider than that of an equivalent stressed skin panel with no internal bonded foam. In a paper by Jorrisen et al at Eindhoven University[9] the method given in the TR019 document[5] and a method in a Skh paper[10], which gave very similar results. The Skh formula is shown below:

$$B_{eff\ (faces)} = \frac{\tanh\left(\frac{\pi}{2} \cdot \frac{(B_p - B_j)}{L} \cdot \sqrt{(E_f/G_f)}\right)}{\left(\frac{\pi}{2} \cdot \frac{1}{L} \cdot \sqrt{(E_f/G_f)}\right)}$$

Where E_f = Young's Modulus of faces
G_f = Shear Modulus of faces

2.5 BENDING (FLEXURAL) STRENGTH CHECKS

Calculation of bending resistance combines the bending strength of the reinforcing joist and the face skins. As it is a composite section, the joist modulus is transformed by the product of the modular ratio of face to joist .

2.5.1 Bending Strength of Panel with Solid Timber Splines (Reinforced)

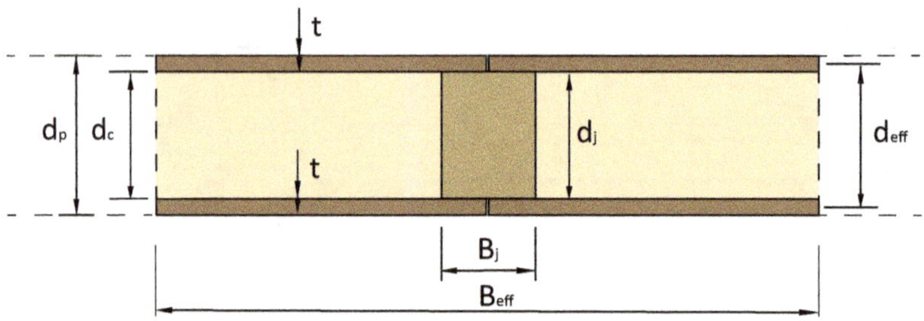

Figure 2.5.1 Local Section (Reinforced)

Thickness of faces = t
Depth of panel = d_p, and Depth of Core = d_c
Effective depth, deff = $d_p - t = (d_p + d_c)/2$

Characteristic resistance in tension of face, $f_{t,0,k}$
Characteristic resistance in tension of face, $f_{c,0,k}$
Noting that the OSB sheet is orientated with outer grain parallel to the span.

Partial factor for material property, Υ_m, where Υ_m OSB/LVL/Kerto = 1.2
Partial factor, Υ_m Joist = 1.3 for solid timber and 1.25 for glulam.
Modification factor for load duration, K_{mod} (assumed service class 2)

(K_{mod} values (OSB): permanent = 0.30, long term = 0.40 and 0.90 instantaneous)
(K_{mod} values (timber): permanent = 0.60, long term = 0.70 and 1.10 instantaneous)
For K_{mod}, service class 1 and 3 (external) refer to Eurocode 5, Table 3.1.
Bending strength of timber, depending on grade/type = $f_{m,k}$

Modular ratio, $M_r = E_j/E_f$....to alter the joist bending strength to a '*transformed*' value.
...for instance if Ej (grade C24, $E_{0,mean}$) = 11,000 N/mm², Mr = 11,000/3800 = 2.89

Design bending resistance in tension, $M_{d,t}$, combining OSB and timber rib. See **figure 2.4.1** for B_{eff}, for a composite section:

$$M_{d,t} = f_{t,0,k} \,.\left(\frac{B_{eff}.\ t.\ K_{mod,f}.\ d_{eff}}{\Upsilon_{m,f}} + M_r.\ \frac{B_j.\ d_j^2.\ k_{mod,j}}{6.\ \Upsilon_{m,j}} \right)$$

for $M_{c,t}$ substitute $f_{c,0,k}$

Panel Properties per 1m Width

For ease of calculation, the properties can be derived for an idealised 1m width panel, where deflection of 1m idealised panel, $d_{max(1m)} = d_{max}.B_p/1000$ in mm and for bending strength $M_{d,t(1m)} = M_{d,t}.1000/B_p$.

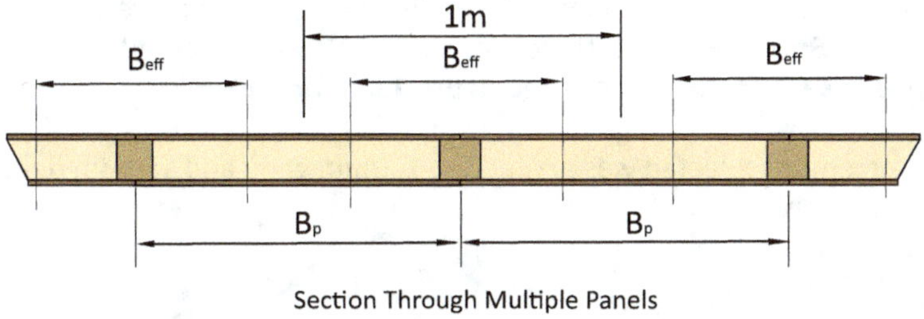

Figure 2.5.2 Idealised 1m Width Panel

2.6 STIFFNESS OF PANEL: FROM BENDING AND SHEAR COMPONENTS

The local section shown in **figure 2.5.1** shows a panel with effective breadth (or width) B_{eff} and with a solid timber spline with depth, d_j, and breadth, B_j.

For a panel with UDL, w, and span L, the deflection midspan may be expressed as:

$$d_{max} = \frac{5.w.L^4}{384.\{E_f.I_f/(1 + k_{def,f}) + E_j.I_j/(1 + k_{def,j})\}} + \frac{w.L^2}{8.A_j.G_j}.(1 + k_{def,j})$$

($d_{max} = d_{max(Gk)B} + d_{max(Gk)V}$) - *sum of bending and shear components). The foam shear deformation is constrained; therefore, this is ignored in the equation.*

E_f = Young's Modulus of skins (faces) - assumed to be 3800 N/mm²
E_j = Young's Modulus of timber splines
$I_f = (d_p^3 - d_c^3).B_{eff}/12$ ....face moment of inertia based on B_{eff}.
$I_j = B_j.d_j^3/12$....joist moment of inertia

assuming compression modulus = tension modulus

Note on modular ratio: the total flexural rigidity (EI) is obtained from the sum of the face rigidity ($E_f.I_f$) and joist rigidity ($E_j.I_j$). It is independent of the modular ratio.

A_j = Area of timber spline ($A_j = B_j.D_j$)
G_j = Shear modulus of timber spline
$K_{def,f}$ = Load duration factor for the face
$K_{def,j}$ = Load duration factor for the timber spline, obtained from **table 2.2.1**.

Calculations for instantaneous, medium term and long term deflections and strength may now be carried out using the combinations and creep coefficients outlined in **section 2.3.2.1** to **2.3.2.3**.

2.6.1 Example Panel Calculations (Timber splines/Reinforced)

175mm Floor Panel in a Garden Office Spanning 3.6m

A garden office, similar to the example in **figure 2.3.3**, was desired by a client with a 3.6m span floor. It was checked with unreinforced panels with SIP splines and was found to deflect significantly under all loading conditions: 17mm (short term), 32mm (medium term) and 25mm (permanent). Reinforced panels were checked using the formulae outlined in **section 2.4** and **2.5**. See **figure 2.6.1** for section.

The timber reinforcement selected was grade C24, 100mm wide and 150mm deep to fit between the 11mm OSB/3 faces with 3mm foam adhesive infill.

Panel width, Bp = 1200mm

Bending strength for C24 timber, $f_{m,k} = 24\ N/mm^2$

Modulus values:

$E_{j,0,mean} = 11{,}000\ N/mm^2$
$G_{j,mean} = 690\ N/mm^2$
$E_{f,0\ osb} = 3800 N/mm^2$
$G_{f,0\ osb} = 1600\ N/mm^2$

$M_r = 11{,}000/3800 = 2.89$

$f_{t,0,k} = 7.7\ N/mm^2$

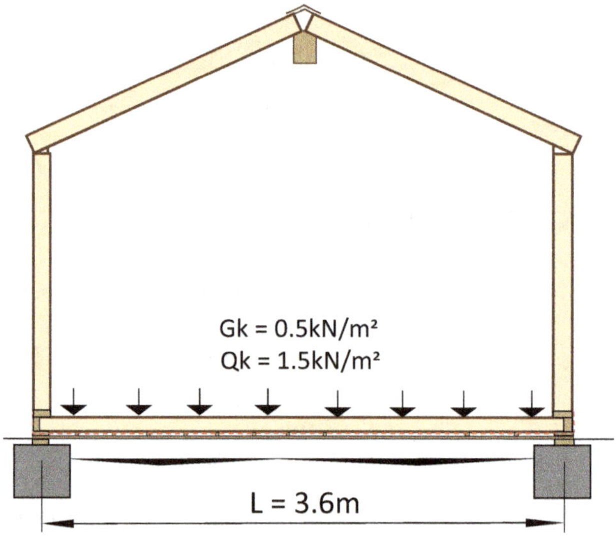

Figure 2.6.1 Garden Office Floor with Reinforced Panels

Calculate B_{eff} from section 2.4.1

$B_{eff} = \tanh(\pi/2 \times (1200-100)/3600 \times \sqrt{(3800/1600)}) / (\pi/2 \times 1/3600 \times \sqrt{(3800/1600)})$
$B_{eff} = \tanh(0.48 \times \sqrt{2.4}) / (0.00044 \times \sqrt{2.4}) = 0.63/0.00068 = 926mm$

Design Bending Resistance in Tension (Long Term)

$$M_{d,t} = f_{t,0,k} \cdot \left(\frac{B_{eff} \cdot t \cdot K_{mod,f} \cdot d_{eff}}{\Upsilon_{m,f}} + M_r \cdot \frac{B_j \cdot d_j^2 \cdot k_{mod,j}}{6 \cdot \Upsilon_{m,j}} \right)$$ ..... from **section 2.5.1**

$M_{d,t} = 7.7 \times 926 \times 11 \times 0.40 \times 164/1.2 + 2.89 \times 7.7 \times 100 \times 150^2 \times 0.70 / (6 \times 1.3)$
$= 4.29 \times 10^6 + 4.49 \times 10^6 = 8.78 \times 10^6\ Nmm \approx 8.8\ kNm$

Assuming long term conditions for live with no reduction:
K_{mod} (OSB/3, service class 2) = 0.40, $K_{mod,j} = 0.70$, $\Upsilon_{m,f} = 1.2$ and $\Upsilon_{m,j} = 1.3$
$d_{eff} = d_p - t = 175 - 11 = 164mm$

Design Bending Resistance per 1m Width

From **figure 2.5.2** for 1.2m wide panels, the equivalent bending resistance per 1m width is given by:

$M_{d,t(1m)}$ = 8.8 x 1000/ 1200 =7.3 kNm/m

Ultimate Bending Moment Applied per 1m Width

Span = 2.5m
Loading, G_k = 0.5 (x1) = 0.5 kN/m and Q_k = 1.5 (x1) = 1.5 kN/m
Ultimate load,W_u = 1.35.G_k + 1.50.Q_k = 1.35x0.5 + 1.50x1.5 = 2.9 kN/m

$M_u = W_u.L^2/8$ = 2.9 x 3.6^2 / 8 = 4.6 kNm/m < 7.3

Note, compared with the method given in **section 2.3.5**, *assuming Beff = 1m:* $M_{d,t(1m)}$ *is 4.6kNm/m for the faces alone and 9.1kNm/m when combined with the bending resistance for the joist. Since Beff < Bp, the calculated value is less. The basis of the assessment is to assume no contribution from the panel section outside the effective breadth (B_{eff}). If the resistance was critical, then it would not be unreasonable to assume* ψ_2 = 0.30 *for floor live loading as discussed in* **section 2.3.2.2**.

Instantaneous Dead Load Deflection

From **section 2.6** deflection is given by the sum of bending and shear components:

$$d_{max} = \frac{5.w.L^4}{384.\{E_f.I_f/(1 + k_{def,f}) + E_j.I_j/(1 + k_{def,j})\}} + \frac{w.L^2}{8.A_j.G_j}.(1 + k_{def,j})$$

If all K_{def} values are zero, for this condition, then:

$I_f = (d_p^3 - d_c^3).B_{eff}/12 = (175^3 - 153^3) \times 928/12 = 137 \times 10^6\ mm^4$
$I_j = B_j.Dj^3/12 = 100 \times 153^3 / 12 = 29.9 \times 10^6\ mm^4$
$A_j = B_j.d_j = 100 \times 153 = 15.3 \times 10^3\ mm^2$

Bending Component (Left Hand Part of Equation)

Face rigidity = $E_f.I_f/(1 + k_{def,f}) = 3800 \times 137\times10^6/(1+0) = 521 \times 10^9\ Nmm^2$
Rib rigidity = $E_j.I_j/(1 + k_{def,j}) = 11{,}000 \times 29.9\times10^6/(1+0) = 329 \times 10^9\ Nmm^2$

Total bending deflection, $d_{max(Gk)B} = 5\times0.5\times3600^4/384/(521+329)/1\times10^9$ = 1.29mm

The equivalent 1m strip deflection, $d_{max(Gk,1m)B}$ = 1.29 x 1200/1000 = 1.54mm

Shear Component (Right Hand Part of Equation)

Shear deflection of rib, $d_{max(Gk)V} = 0.5 \times 3600^2/8/15.3 \times 10^3/690$ = 0.08mm
The equivalent 1m strip deflection, $d_{max(Gk,1m)B}$ = 0.08 x 1200/1000 = 0.09mm

The contribution is less than 5% of the total deflection, having much lower influence than foam shear deformation in an unreinforced panel.

Instantaneous Live Load Deflection

Substituting live loading into the equations above the central deflection was obtained:

$d_{max(Qk)B}$ = 3.8mm, $d_{max(Qk,1m)B}$ = 4.6mm
$d_{max(Qk)V}$ = 0.23mm, $d_{max(Qk,1m)V}$ = 0.28mm

Medium Term and Permanent Dead and Live Load Deflections

Deflections were calculated based on creep factors used in **section 2.3.5** and applied to the previous formulae. **Table 2.6.1** summarises the ***1m equivalent*** deflections.

Material (1m width) 1200mm rib spacing	Instantaneous	Medium Term 2000 hours (84 days)	Permanent 220k+ hrs (25+ years)
Dead $d_{max(Gk)faceB}$	1.5mm	2.7mm	3.9mm
Dead $d_{max(Gk)ribV}$	0.1mm	0.1mm	0
Live $d_{max(Qk)faceB}$	4.6mm	6.3mm	0
Live $d_{max(Qk)ribV}$	0.3mm	0.3mm	0
Total Deflection	6.5mm	9.4mm	3.9mm

Table 2.6.1 Summary of Deflections (175mm reinforced floor panels, 3.6m span)

For example, $d_{max(Gk)faceB}$ (medium term) by applying $k_{def,f}$ = 1.1 and $k_{def,j}$ = 0.4
= $0.5 \times 5 \times 3600^4/384/(521/(1+1.1)+329/(1+0.4))/1 \times 10^9$ = 2.3mm

Equivalent Im width deflection = 1200/1000 x 2.3 = 2.7mm

Effect of Closing Up Rib Spacing with 600mm wide Panels

The panel may be stiffened by halving the width of each panel to 600mm and installing timber splines in each joint; calculation results are summarised below:

B_{eff} = 479mm - *note for **1m strip** multiply values by 1000/600 (1.67)*

Material (1m width) 600mm rib spacing	Instantaneous	Medium Term 2000 hours (84 days)	Permanent 220k+ hrs (25+ years)
Dead $d_{max(Gk)faceB}$	1.1mm	1.8mm	2.5mm
Dead $d_{max(Gk)ribV}$	0.1mm	0.1mm	0
Live $d_{max(Qk)faceB}$	3.3mm	4.3mm	0
Live $d_{max(Qk)ribV}$	0.1mm	0.2mm	0
Total Deflection	4.6mm	6.4mm	2.5mm

Table 2.6.2 Summary of Deflections for 600mm Wide Panels (175mm, 3.6m span)

Summary of Garden Office 3.6m Span SIP Panel (Reinforced) Deflections

A summary of the results in **table 2.6.1** (ribs at 1200mm centres) and **table 2.6.2** (ribs at 600mm centres), for deflections based on K_{def} values from **section 2.3**, and moment resistance calculation follows:

- Strength is not an issue. The resistance $M_{d,t}$ = 7.3kNm/m, is greater than the applied design moment of 4.6kNm/m.
- The instantaneous and medium term deflection is by ratio to span, respectively L/553 (6.5mm) and L/383 (9.4mm), which is higher than the L/250 criterion.
- The fundamental natural frequency, based on instantaneous G_K+10%Q_K loading (2.1mm) and the formula from the previous garden office example is 12.2Hz > 8.
- In the short term the deflection is greater than the permanent sag. This contrasts with the **table 2.3.1** figures whereby permanent sag was much lower, demonstrating the influence of the core shear creep in unreinforced panels.
- The rib shear deformation, $dmax_{(Qk)ribV}$, contributes less than 5% of the total deflection. Its influence increases for shorter spans.
- Deflections for an equivalent panel, unreinforced are: 17mm (short term), 32mm (medium term) and 25mm (permanent). The difference is most marked for permanent sag, which demonstrates that the addition of ribs at 1.2m and 0.6m centres has a very significant stiffening effect.
- Panels with 600mm width, with splines in each joint, have a greater stiffness than the 1200mm wide panels by a margin of circa 40-60% for all durations of loading.

2.7 IDEALISED SECTION PROPERTIES FOR MODELLING PLATE ELEMENTS

Finite element models (FEM) with 2D plate elements are very useful for assessing panels with three or four sided supports with two way spanning action and for complex geometry, such as with a pitched roof with hips, valleys and dormers.

Figure 2.7.1 shows an idealised panel which is assumed to be a homogeneous material, with depth, d_p, and breadth, B_p. The moment of inertia, I_p, is obtained by calculation assuming a rectangular homogeneous section, and Young's modulus, E, is found from the total deflection found from UDL analysis in the previous section:

$$d_{max} = \frac{5.w.L^4}{384.E_p I_p} \quad \text{.... therefore, } E_p = \frac{5.w.L^4}{384.\ d_{max}.I_p}$$

For example, taking the long term sag for the 142 panel in **Table 2.3.1**, of 40.5mm, and dead load (no live for long term), G_{ks} = 1.04, and Ls = 2.9m:

$$I_{p(1m)} = 1000\text{x}142^3/12 = 239\text{x}10^6 \text{ mm}^4$$

$$E_p = 5\text{x}1.04\text{x}2900^4/384/40.5/239\text{x}10^6$$
$$= 99 \text{ N/mm}^2$$

$$I_p = \frac{d_p.B_p^3}{12}$$

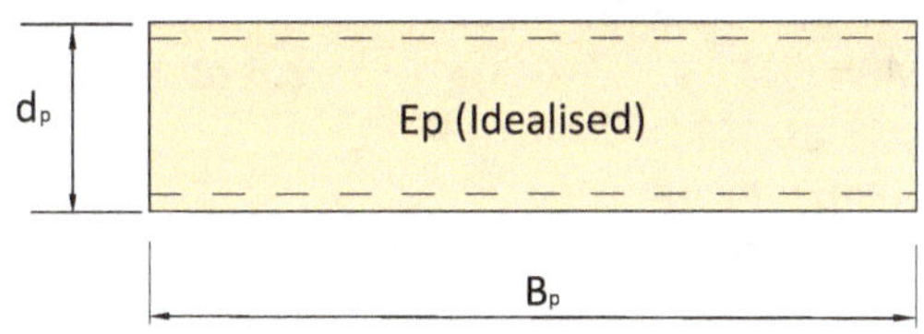

Figure 2.7.1 Idealised Homogeneous Section

2.8 FINITE ELEMENT MODEL DEMONSTRATING 2 WAY SPAN ACTION

The benefits of two way spanning panels will be discussed in this section, with a note of caution that the span action perpendicular to the short span should be represented by a reduced stiffness, as summarised in **section 2.3.1**. A conservative approach has been taken in this exercise to assume that the secondary span has a stiffness that is 10% of the main (short) span.

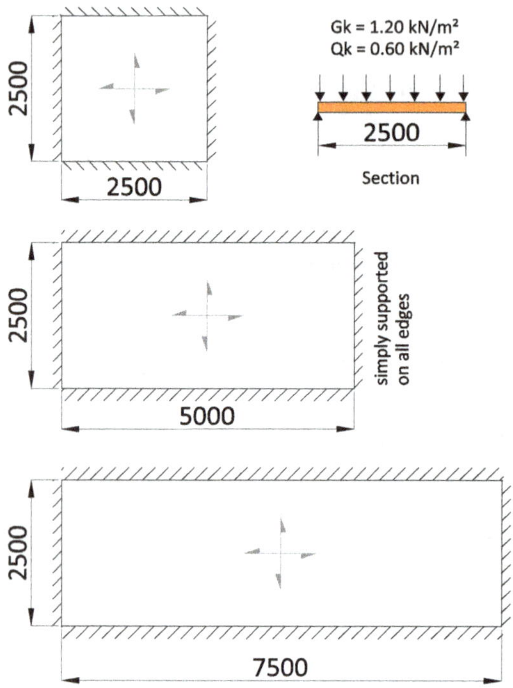

Figure 2.8.1 Two Way Span Panels

Figure 2.8.1 shows three 125mm roof layouts, which have an increased aspect ratio of long span to short span, from 1 to 3.

The panels are horizontal and have loading shown.

They are supported on all four edges with no moment restraint assumed.

The panels were modelled with idealised sections, as described in **section 2.7**, with plate elements that have "elastic orthotropic properties". In other words, with moduli which differ in orthogonal directions to allow for reduced stiffness in the secondary span.

The check assumed long term values for rigidity, since roof sag is critical to the design.

Stage 1: Derive Long Term Idealised Modulus Under Dead Load Only

Based on calculation methods described in **section 2.3**, *ignoring the edge supports* and assuming a span of 2.5m, the long term deflection and ultimate short term bending moments are shown below for the 125mm panel with 11mm OSB faces:

$d_{max(Gk)lt}$ = 38.7mm (dead load only)
M_u = (1.2 x 1.35 + 0.60 x 1.5) x $2.5^2/8$ = 2.0kNm/m (< resistance 4.0kNm/m)

Span, L = 2.5m
Loading perpendicular to panel, G_k = 1.2kN/m, Q_k = 0.6kN/m

Idealised moment of inertia, $I_p = 125^3 \text{x} 1000/12 = 163 \text{x} 10^6 \text{ mm}^4$
Idealised modulus, $E_p = 5 \text{ x } 1.2 \text{ x } 2500^4/384/38.7/163 \text{ x } 10^6 = 97 \text{ N/mm}^2$ (*95 used*)

Stage 2: Model Panels with 2D Plate Elements

Based on calculation methods described in **section 2.3**, *ignoring the edge supports* and assuming a span of 2.5m, the long term deflection and ultimate short term bending moments were output.

Mesh and Plate Elements

The model was run with a mesh of 8x8 elements. The software used was GSA by Oasys, which has the facility for orthotropic elements. This may not be available on other software. The main properties of plate elements are shown in **table 2.8.1**.

2 Way Span Model Summary of Properties	
Shell Element Thickness	125mm
In Plane Shear Factor	1% (v.low)
Bending Factor	100%
Ex Modulus N/mm²	10 (c.10%)
Ey Modulus N/mm²	95 (100%)
Ez Modulus N/mm²	10 (v.low)
Poisson's Ratio	0.15

Table 2.8.1 Element Properties

In plane shear was conservatively set to a low value to minimise in plane shears assisting in load bearing. It has been found in general models that setting this to 30-35% gives reasonable results.

The 'X' direction is parallel to the secondary span. 'Ex' is circa 10% of the primary.

Primary span modulus was calculated previously.

'Ez', the modulus in the Z direction was minimised.

Calibration Model

Calibration Model
Maximum deflection
38.7mm

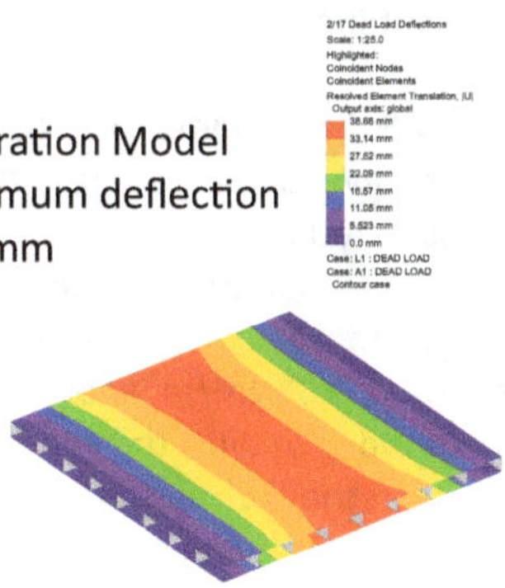

Figure 2.8.2 Calibration Model Deflections

An initial model was run with the edges unsupported on the slope to check results against the initial calculations.

This is recommended as a check.

Figure 2.8.2 gives output for deflections under dead loading for the calibration model. The maximum deflection at 38.7mm corresponds closely to the calculated deflection in the previous section.

If the value is not close, trial and error adjustment of 'Ey' can be undertaken once the model is set up. As long as the model has sufficiently small meshing and the dimensions and loading correspond to the roof this should yield consistent results when the edge supports are added.

2.5x2.5m Roof Model Output

The edge supports were added along the slope and the model was run. **Figure 2.8.3** shows deflection output.

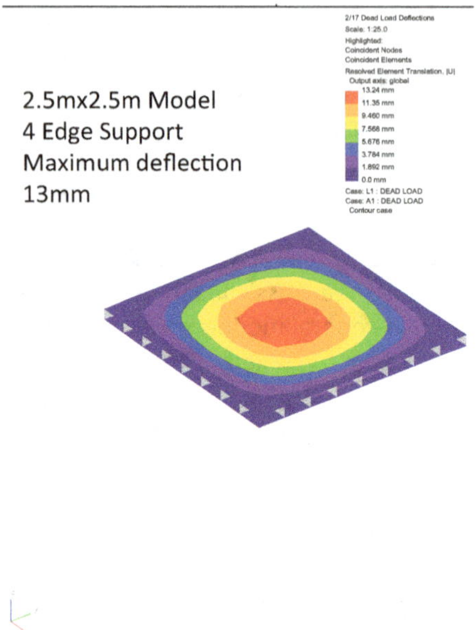

Figure 2.8.3 2.5m x 2.5m Model Deflections

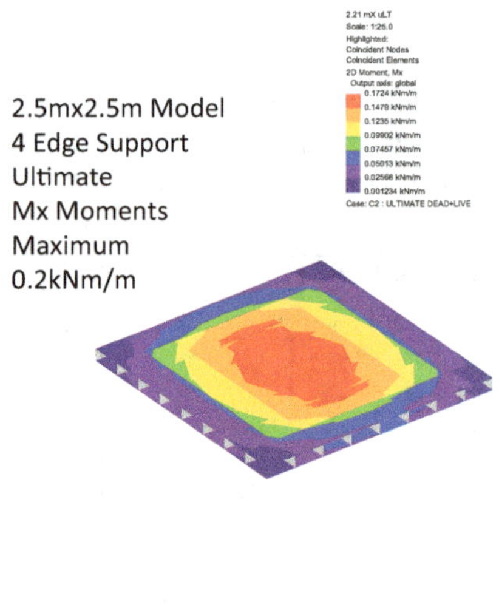

Figure 2.8.4 2.5m x 2.5m Model Mx Ultimate

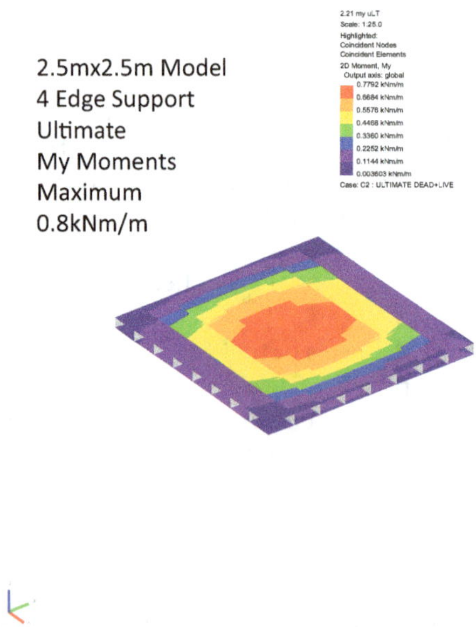

Figure 2.8.5 2.5mx2.5m Model My Ultimate

Maximum deflection at 13mm is circa 30% less than for the analysis assuming one way spanning panels. This is a significant increase in stiffness.

The bending moments for *ultimate dead and live* loading are less than the calculated moment of 2.0kNm/m. Mx in the secondary direction is 0.20kNm/m. My in the primary direction is 0.8kNm/m, showing correspondence with the reduced stiffness in the X direction.

In reality, the 2.5m width would need to be split into three panels. To save waste, that might be a 1.2m panel centrally with two 0.65m wide panels alongside, running up the slope with three sided support, giving a marginally stiffer arrangement.

5.0x2.5m Roof Model Output

Results for the 5m wide panel are shown below.

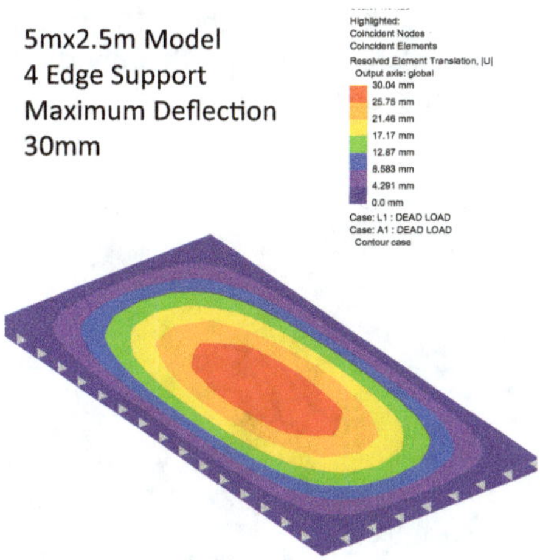

Figure 2.8.6 5.0m x 2.5m Model Deflections

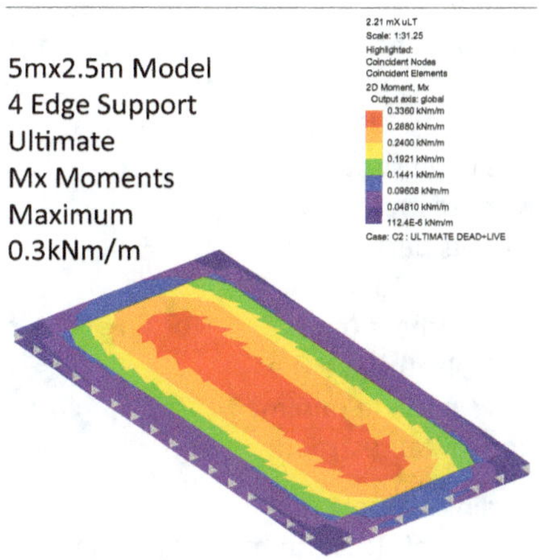

Figure 2.8.7 5.0m x 2.5m Model Mx Ultimate

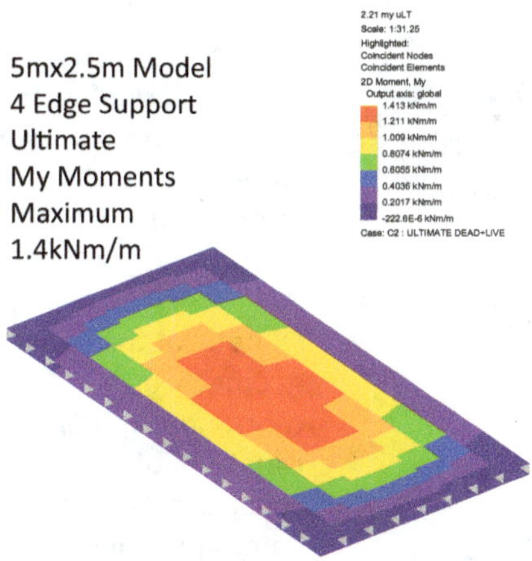

Figure 2.8.8 5.0m x 2.5m Model My Ultimate

In summary:

$dmax_{(Gk)lt}$ = 30mm
$M_{x(ult)}$ = 0.3kNm/m
$M_{y(ult)}$ = 1.4kNm/m

For a panel that is twice as wide as the main span, the deflection, and primary span direction moment (My) are both approximately 75% of My for 1 way spanning panel with edges unsupported. The two way span stiffening effect is of greatest influence for width to span aspect ratios in the range 1 - 1.5.

7.0x2.5m Roof Model Output

For a panel which is three times as wide as the span, results are summarised below:

$dmax_{(Gk)lt}$ = 37mm
$M_{x(ult)}$ = 0.3kNm/m
$M_{y(ult)}$ = 1.8kNm/m
Deflection and bending in the primary direction are both close to the one way span.

2.8.1 Finite Element Model of a Roof

Figures 2.8.9 and **2.8.10** show a complex roof with dormers which was modelled to demonstrate deflections for a tiled roof built with 142 Kingspan panels (15mm OSB faces) with SIP splines. The mesh was formed with plate elements with properties derived as described in the previous section, a combination of 'quad' and 'triangular' elements. The maximum dead load permanent deflection is 12.9mm.

The edge supports were restrained with springs perpendicular to walls with a low stiffness (circa 100kN/m) since the walls provide minimal horizontal restraint.

Purlins support the main roof. This can be seen in **figure 2.8.10** as beam elements at the top of the dormers. There are intermediate supports from load bearing walls.

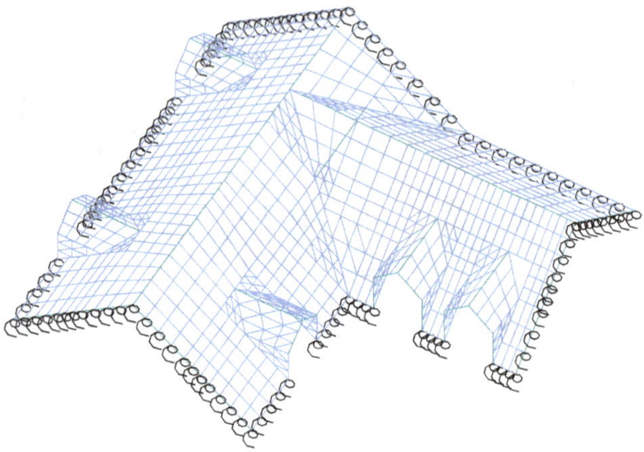

Figure 2.8.9 Complex Roof Model - Mesh (Plate Elements)

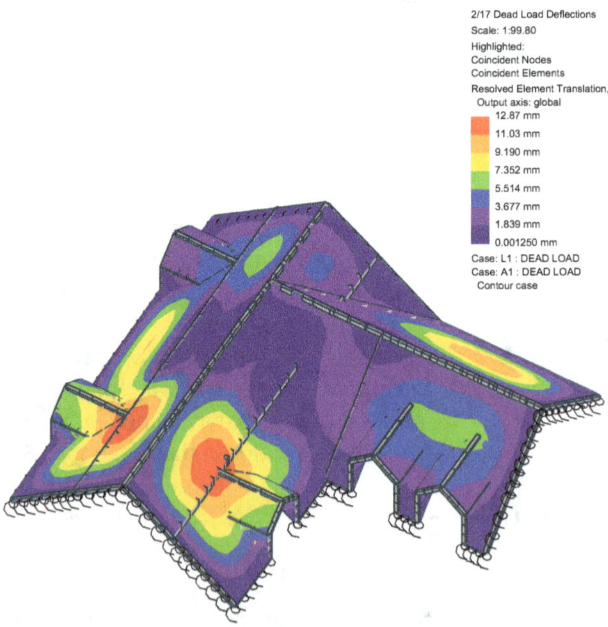

Figure 2.8.10 Complex Roof Model Dead Load Deflections Long Term

Perpendicular to the main roof, reinforced panels with solid splines were incorporated so as to omit purlins in this area.

Of note, at the folded geometry of the roof, and diaphragm capability of SIP construction of the panels provides rigidity as the apex and in the valleys.

Dormers also stiffen roof panels in a similar way, as can be seen for the reinforced roof area, where they sit on the right hand side of the roof.

2.9 MULTIPLE SPAN PANELS - CHEQUERBOARD LAYOUT

SIP panel roofs are often employed for flat roofs, for instance, in commercial projects or schools with regular supports at circa 3-5m apart. If the panels are arranged in a manner whereby the end to end joints are placed at the points of contraflexure this helps to minimise deflections, in particular for uniform, long term dead loading.

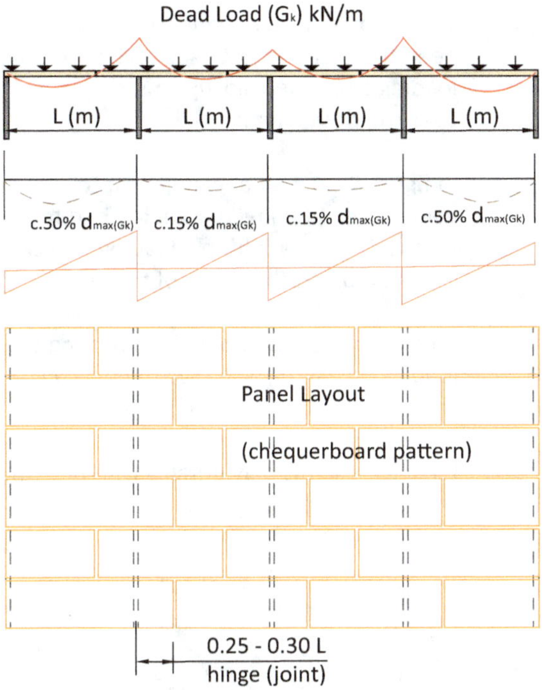

Figure 2.9.1 Multiple Span Roof Panel Layout

In **figure 2.9.1** a four equal span roof has a constant dead loading over its extent.

As noted in previous examples for panels with SIP or timber splines, roof design is mostly determined by medium to long term deflections; not strength.

By arranging the panels in a chequerboard layout - with end joints located at the quarter to third span positions (at the point of contraflexure for continuous spans) this results in deflections, which are circa 50% of the equivalent simply supported span deflections[5].

Where the panels are joined end to end, there is no need to consider transfer of bending across the joint since they connect at hinge points.

Furthermore, if panels are set out in chequerboard fashion, as discussed in **section 3.3** for floor panels on joists, this is the most robust arrangement for providing diaphragm resistance to lateral loads.

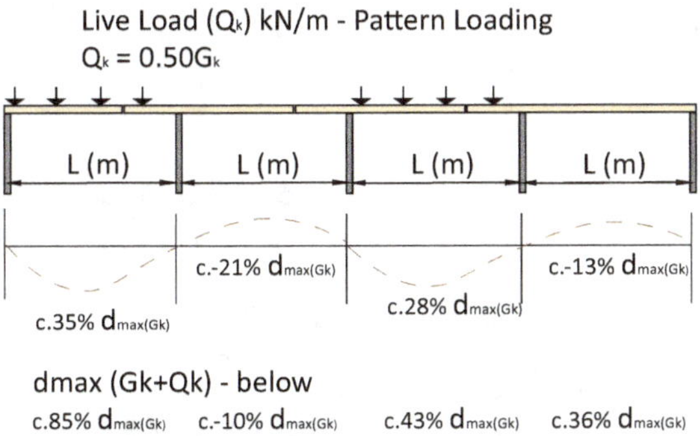

Figure 2.9.2 Pattern Live Loading (Q_k = 50% G_k)

Medium and short term checks should allow for live pattern loading. In **figure 2.9.2,** with live loads assumed to be 50% of dead loads, the alternate loading pattern, combined with dead load deflections are a maximum of circa 85% of equivalent simply supported dead load deflection.

Loads from mechanical plant may also require additional supports locally.

2.10 VOIDS IN PANEL LAYOUTS

When voids are formed in roof or floor panels for rooflights or stair wells, for instance, panels need to be arranged around the voids with timber trimmers or steel, if necessary, designed to support the panels locally. See **figure 2.10.1** and **2.10.2,** showing a flat roof and pitched roof trimming arrangement.

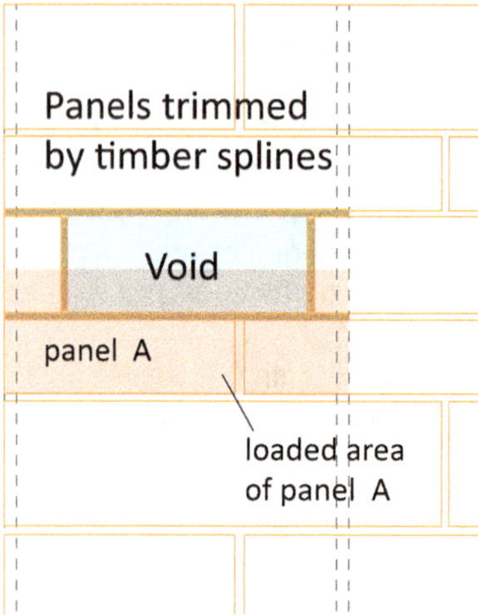

Figure 2.10.1 Flat Roof Void Trimmers

To support the void, panels flanking the void need to be designed to support additional loading from the void without deflecting over the design limits.

In **figure 2.10.1** a panel adjacent to the void - labelled panel A - would need to be checked for additional loading, which equates to half the width of the void.

If the panel is incapable of supporting the load the trimmer, will have to be designed to support this additional line load.

If necessary this may entail a trimmer which is deeper than the SIP panels.

In a large rooflight void within **figure 2.10.2** a pitched roof over a stairwell glulam trimmers, a and b, support the edges. Deeper trimmers, c and d running the length of the slope, are supported off a glulam ridge beam, and, at the lower end, the external SIP wall at eaves level.

Beams **a** and **b**, which carry SIP panels and glazed frame, are checked with UDL loads derived from the contributory area found from half panel widths. The worst case is to assume 1.2m wide panels adjacent to the trimmers.

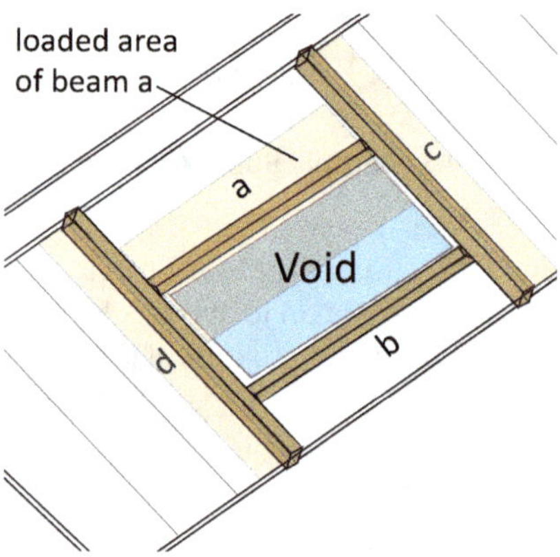

Figure 2.10.2 Pitched Roof Large Void

Beams **c** and **d** are checked with a combination of the point loads from supporting beams **a** and **b**, and a UDL strip load with contributory width assigned from half the width of the adjacent SIP roof panel. The trimmers may be deeper than the SIP panels, visible from the underside of the roof, and would need to be advised to the client and architect for their acceptance. To reduce depth the trimmers can be built into the SIP panels.

2.11 SHEAR AT PANEL SUPPORTS

The shear strength of panels is not simply related to the depth of the panel. According to Tompos[8] overall shear resistance reduces with an increase in depth of a panel. A plot for depths from 110mm (4.5") to 320mm (12.5") gave ultimate shear strengths of 0.11N/mm² and 0.05N/mm² respectively: equivalent to 12.5kN/m and 15kN/m. This is assigned a 'C_v' value in the reference.

Rungthonkit[6] summarised shear tests to failure for 4 point loaded (600mm wide, 1.1m span) 125mm SBS panels on a bearing support for three types: no central spline (STP), central SIP (SMC) spline and central 47mm wide timber spline (SDC). Shear failure for the total load was 19.3kN minimum for STP and SMC panels and 40.3kN for SDC. The STP and SMC panels failed at the supports and the SDC under the third span load in flexural shear. Over 1m width, and assuming panels STP and SMC are the same, this gives, characteristic values of:

STP/SMC characteristic F_{vc} = 19.3/2 x 1/0.6 = 16kN/m
SDC characteristic, $F_{vc,rein}$ = 40.3/2 x 1/0.6 = 34kN/m

Assuming that deeper panels will have no greater resistance in shear (see Tompos[8]) for long term failure (say K_{mod} = 0.50, and Υ_m = 1.25 based on core modification factors in **section 2.12**), then all OSB/3 SIPs shear capacities are as follows:

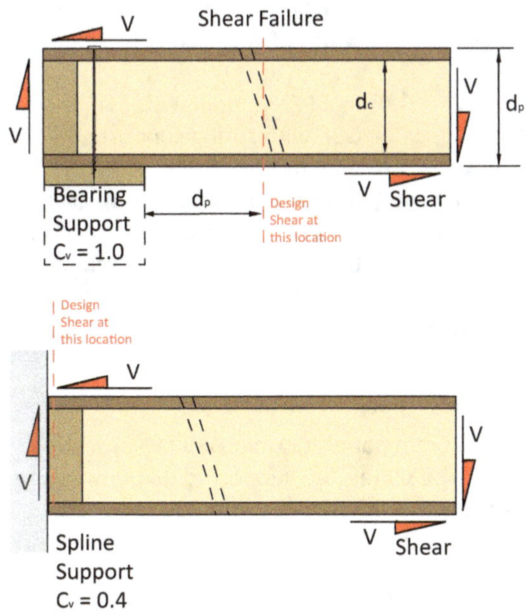

Figure 2.11.1 Shear Failure at Support

Bearing Support
(Long Term Shear Strength)

F_{vd} = 16 x 0.50/1.25 = 6.4kN/m
$F_{vd,rein}$ = 34 x 0.50/1.25 = 13.6kN/m

Spline Support

C_v = 0.40 in **figure 2.11.1**

F_{vd} = 0.40 x 6.4 = 2.6kN/m
$F_{vd,rein}$ = 0.40 x 13.6 = 5.4kN/m

Shear resistance for panels is rarely a concern as deflection is generally critical.

For example, a 3.5m span green roof panel with high loads of 1.5kN/m² dead and 0.75kN/m² live load equates to a support ultimate shear of, V = 5.5kN/m. *This span and loading would necessitate timber splines:*

V_d = 13.6kN/m > 5.5

2.12 LOCAL POINT LOADS

If panels bear on supports or have a point load with a small bearing area acting transversely, with no timber spline at the end of the panel, the resistance of panels to resist the loads is limited by the core in compression. An example is illustrated in **figure 2.9.1**, with panels spanning onto intermediate walls in a multiple span layout. Timber splines increase the panel resistance.

Bearing Method

This is a complex method which takes into account a combination of compressive stresses in the base skin in a continuous support. See **figure 2.12.1,** where the bottom face is subject to 'wrinkling' under compression and bearing compression under reaction R. A calculation, based on TR019[5] section 5.2, was carried out for 11mm OSB bottom skin to assess the skin wrinkling (or local bending effect). This was found to be negligible under typical panel loading.

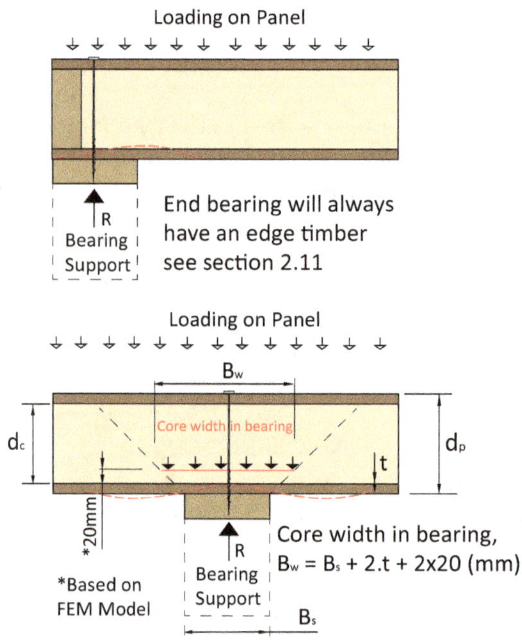

Figure 2.12.1 Bearing Support Core Compression

Resistance to the panel loading can be assessed primarily by checking the core compression under load. In **figure 2.12.1** the bearing area for an end and internal condition of loading is found by obtaining the core width in bearing, B_w, under which the reaction load is spread. Owing to the flexure of the base skin, which is analogous to a beam on 'elastic foundations,' the effective width of core acting is increased.

In a FEM model with 11mm OSB with spring supports the effective bearing width was found to be circa 50mm, giving similar compressive stresses but this was reduced to 20mm for conservatism.

Calculate Design Compression Resistance (per 1m) in Core

Assuming a multi span (L = 2.2m) floor of 175mm panels (11mm OSB faces) with loading, $G_k = 0.5kN/m^2$ and $Q_k = 1.5kN/m^2$. The panels are supported on 90mm wide studs. Reaction at the first internal support is the worst case value, assumed to be 20% greater than equivalent simply supported condition:

Ultimate load, $w_u = 0.5 \times 1.35 + 1.5 \times 1.5 = 2.9kN/m^2$
Reaction at first internal support, $R_1 = 1.2.w_u.L = 1.2 \times 2.9 \times 2.2 = 7.6kN/m$ >> 2.1
Effective bearing width, $B_w = B_s + 2.t + 2 \times 20 = (90 + 2 \times 11 + 2x20) = 152mm$
End *core area in Bearing/m*, $A_{cb} = B_w \times 1000 = 152x10^3\ mm^2$

Compression Strength of Core[6]

Compression tests on a PUR core from Rungthonkit[6] demonstrated:
Compressive strength, $f_{cc,PUR}$ = 0.25 N/mm² when sandwiched between OSB sheets.

For EPS sheets used in Eco SIPs EPS panels, compressive strength, $f_{cc,eps}$ = 0.10 N/mm²

Modification Factors for Duration and Material for Core

From TR019[5] E.3.2.2 recommends K_{mod} factors for service class 1 and 2 are:

Instantaneous	1.00
Medium term	0.75
Long term	0.50
Permanent	0.25

Material factor for core, Υ_{mc} = 1.25

Core Compression Design Resistance Calculation - Internal Support

Assuming long term loading

Live at 1.5kN/m² is assumed to have a relatively short duration:

Design resistance of PUR, $F_{cd,PUR} = f_{cc,PUR}.K_{mod}/\Upsilon_{mc}$ = 0.25 x0.50/1.25 = 0.10 N/mm²
Equivalent to a resistance line load, $P_{ld,PUR} = F_{cd,PUR}.B_w$ = 0.10 x 152 = 15.2kN/m > 7.6

Design resistance of EPS, $F_{cd,EPS} = f_{cc,EPS}.K_{mod}/\Upsilon_{mc}$ = 0.10 x 0.50/1.25 = 0.040 N/mm²
Equivalent to a resistance line load, $P_{ld,EPS} = F_{c,PUR}.B_w$ = 0.040 x152 = 6.1kN/m < 7.6

Note if *variable action factor* ψ_2 = 0.30 for live, from **figure 2.3.3** example:
R_2 = (0.5x1.35+0.3x1.5x1.5) x 1.2 x 2.2 = 3.6kN/m < 6.1 - *EPS core is adequate*

Assuming permanent (dead load only) condition

Design dead load, R_{Gkd} = 1.2 x (0.5 x 1.35) x 2.2= 1.8kN/m

Resistance line load, $P_{ld,PUR(\psi 0.3)}$ = (0.25 x 0.25/1.25) x 152 = 7.6kN/m > 1.8
Resistance line load, $P_{ld,EPS90\,(\psi 0.3)}$ = (0.10 x 0.25/1.25) x 152 = 3.0kN/m < 1.8

Comparison of Shear Resistance and Bearing Resistance

In **section 2.11,** support shear resistance for PUR core panels with no timber splines was calculated to be 6.4kN/m for long term loading. The bearing resistance from calculation based on **section 2.12** for the end support, long term with PUR foam core, is 15.2kN/m. From BBA certificates quoted internal support long term values are:

- 142mm Kingspan TEK (PUR core) Shear: 7.62kN/m and Bearing: 7.3kN/m
- 144 Eco SIPs (EPS core): Shear: 3.33kN/m and Bearing: 5.09kN/m

The end shear condition gives a reasonable comparison, but bearing calculations for PUR foam are somewhat greater than the BBA values for Kingspan TEK panels. If BBA values are available, the reader is advised to use the quoted figures. In the absence of BBA bearing values for a particular panel assumption of 100 kN/m² (0.10 N/mm²) characteristic compressive strength for PUR foams is recommended, although bearing and shear are rarely critical for most SIP panels on roofs or floors.

2.12.1 'Punching Shear' Under Point Loads

In **figure 2.12.2**, a timber post sits on a SIP panel supporting a raised deck. The critical failure is the compression of the core in bearing. In a similar manner to the previous support condition, the core area under compression is calculated assuming that its load is spread over an extended area under the post. Using the compression resistances, and load spread distribution, from the previous section, a maximum point load, $P_{pd,PUR/EPS}$, which may be supported, is calculated:

Post dimensions, B_{sp} = 100mm and D_{sp} = 100mm
Thickness of face, t = 15mm

From previous page, PUR core maximum ultimate post load carried (*long term* condition, $F_{cd,PUR} = 0.10\ N/mm^2$ and $F_{cd,EPS} = 0.04\ N/mm^2$)

Bearing area,
$A_{pb} = (100 + 2 \times 15 + 2 \times 50)^2$
$= 52{,}900mm^2$

$P_{pd,PUR} = F_{cd,PUR} . A_{pb}$
$= 0.10 \times 52{,}900/1 \times 10^3$
= 5.3kN (≈ 3.6kN *unfactored*)

Similarly, $P_{pd,EPS}$ = 2.2kN
(≈ 1.5kN *unfactored*)

Clearly point load resistances are limited. Increased strength may be achieved with timber splines located under posts, or load spreader plates (timber or other rigid material).

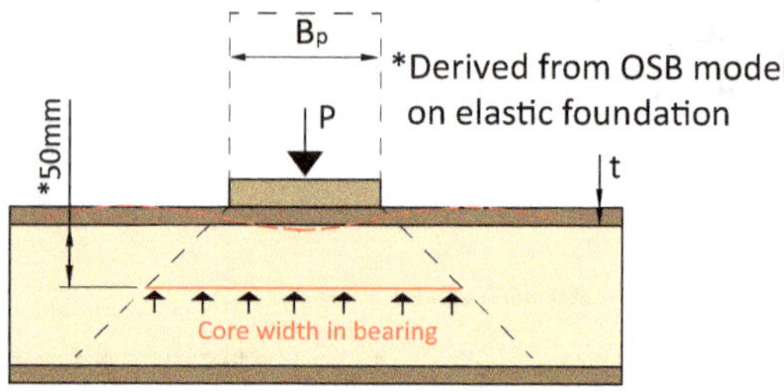

Core width in bearing,
$B_w = B_p + 2.t + 2 \times 50$ (mm)
$D_w = D_p + 2.t + 2 \times 50$ (mm)

Substitute B_{sp} and D_{sp} for B_p and D_p respectively

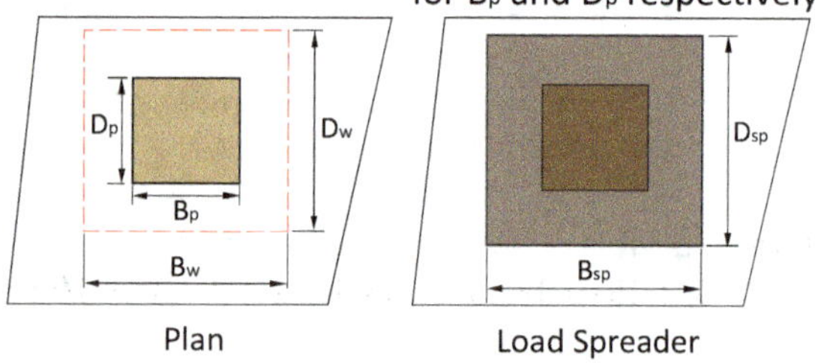

Figure 2.12.2 Bearing Under Point Load

Screw Fixing Uplift Failure

SIP roof panels are generally fixed into supporting walls, lintels and beams with flange headed screws that are sufficiently long to extend through panels, into the supporting header or beam below. In **figure 2.12.3** three critical failure mechanisms are shown, where a roof panel is subjected to suction or uplift loads:

- Core bearing failure.
- Head 'pull-through' (screw head tearing through OSB face).
- Screw withdrawal out of the support beam or header.

An example is evaluated overleaf.

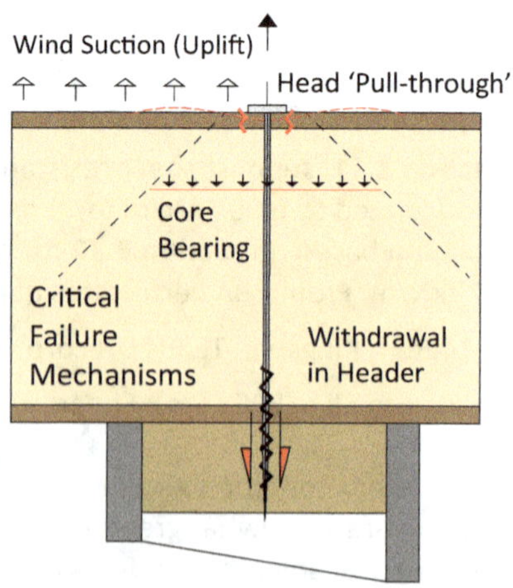

Figure 2.12.3 Uplift - Critical Failure Mechanisms

Uplift Fixing: Core Bearing Failure

If the local wind suction causes a net uplift, a mode of failure is core compression in a similar manner to the example of a panel sitting on a wall. A 300mmx300mm finite element model (FEM) was run, which was represented by plate elements 11mm thick meshed OSB properties with 33mm subdivisions, each supported on a spring supported node, with a stiffness set to the modulus of the core at 8.0N/mm². The criterion for failure was a 2-3mm maximum distortion based on Tompos[8] beam on elastic foundation formula assuming worst case deflection 1/8". See **figure 1.12.5**.

It was found that the maximum distortion for a 4kN central load gave a maximum equivalent pressure on the supporting core of 0.19N/mm². The area of influence based on distortion was bounded by a circular area with a diameter approximating to 150mm.

This tied in with the load spread of 50mm depth for a 45° load spread, which was assumed for the panel on wall check.

A calculation for core resistance based on parameters in **figure 2.12.4** follows:

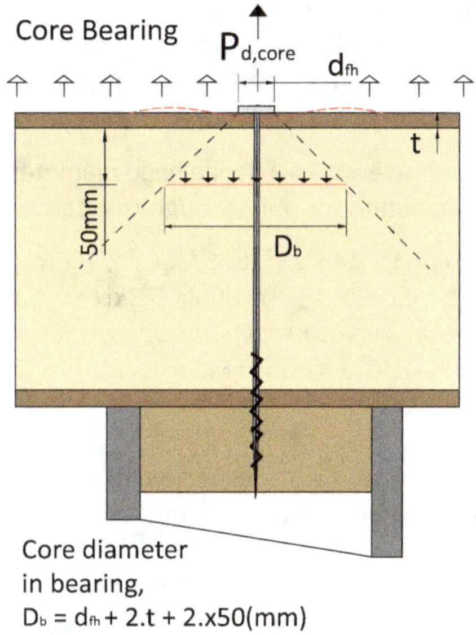

Figure 2.12.4 Core Bearing Failure Mode

Screw uplift resistance, $P_{u,core}$, found from a core bearing circular area with diameter D_b, under flange headed screw head diameter, d_{fh}, of 15mm. Face thickness, t_1 = 11mm.

Bearing diameter, $D_b = d_{fh} + 2.t + 2.50$
$= 15 + 2 \times 11 + 100 = 137$mm
$A_{pb} = \pi.d_b^2/4 = \pi \times 137^2 / 4 = 14{,}741$mm²

Resistance of PUR, $F_{c,PUR}$ - *instantaneous*
$F_{cd,PUR} = f_{cc,PUR}.K_{mod}/\Upsilon_{mc} = 1.0 \times 0.25/1.25$
$= 0.20$ N/mm²

Similarly, for EPS, $F_{cd,EPS} = 0.08$ N/mm²

Therefore:
$P_{pd,PUR} = A_{pb}.F_{cd,PUR} = 14{,}741 \times 0.20/1000$
$P_{pd,PUR}$ = **2.9kN** < 4 (*above, conservative*)
$P_{pd,EPS} = A_{pb}.F_{cd,EPS}$ = **1.1kN**

For screws at 350mm spacing, for example, this gives an equivalent line load resistance of 8.2kN/m (2.9/.35) for PUR foam and 3.1kN/m for EPS foam.

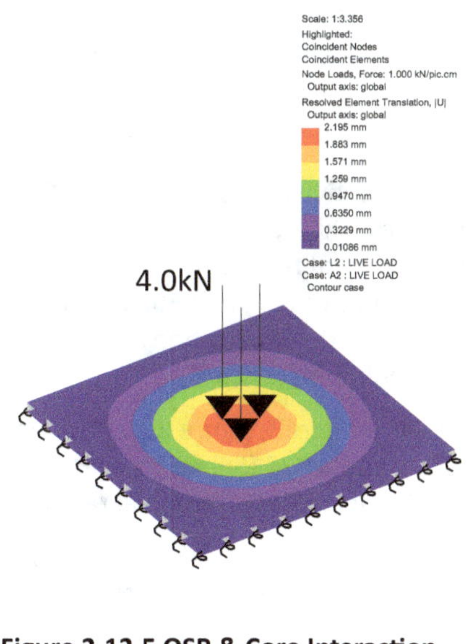

Figure 2.12.5 OSB & Core Interaction Model

Uplift Fixing: Pull Through Failure

BS EN 1995-1.1 (8.40b) provides a formula for pull-through characteristic screw resistance:

$P_{d,PT} = n_{ef}.f_{head,k}.d_{fh}^{2}.(\rho/350)^{0.8}$
where...

$f_{head,k}$ = pull through material face resistance *(14 N/mm² for OSB with flange head screw)*
n_{ef} = number of screws
d_{fh} = screw flange head diameter
ρ = face density (665 N/mm³ OSB)
Service class 2 is assumed

For a screw with a flange head diameter 15mm, calculation for pull through resistance, $P_{u,PT}$:

$P_{d,PT} = 1 \times 14 \times 15^{2}.(665/350)^{0.8}/1000$
$= 14 \times 225 \times 1.9^{0.8}/1000 = 5.3kN$
Design resistance, instantaneous (OSB):
$= P_{d,PT}.K_{mod}/\Upsilon_{m(OSB)} = 5.3 \times 1.1/1.25$
= **4.6kN** > 2.9 - *core bearing is critical*

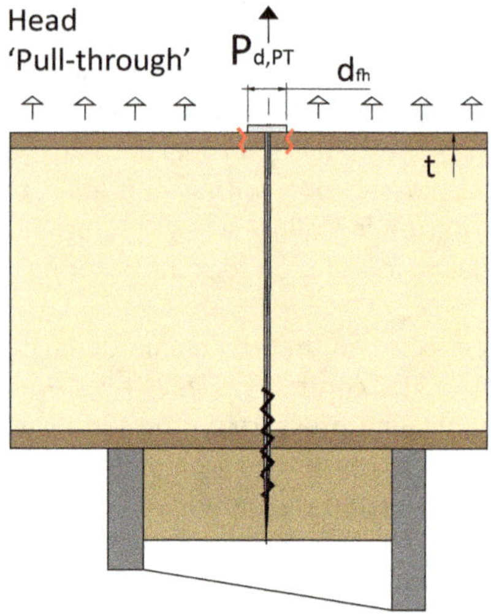

Figure 2.12.6 Pull Through Failure Mode

Note that this is irrespective of core type, and $f_{head,k}$ *is based on a range of face thicknesses, t = 12 - 20mm.*

Uplift Fixing: Withdrawal Failure

Eurocode 5 (8.38 & 8.39) provides a formula for characteristic withdrawal resistance of a screw:
$f_{ax,90,Rk} = \eta_{eff}.f_{ax,k}.d.l_{eff}.K_{d}.$
assuming screw pullout is perpendicular to the grain, where...

$f_{ax,k} = 0.52.d^{-0.5}.l_{eff}^{-0.1}.\rho^{0.8}$ *...in N/mm²*

η_{eff} = effective number of screws ``
d = screw shaft diameter
ρ = density of timber holding thread (310 kg/mm³: C16, and 350 kg/m³: C24)
l_{eff} = pointside penetration
K_d = minimum: (1.0,d/8)

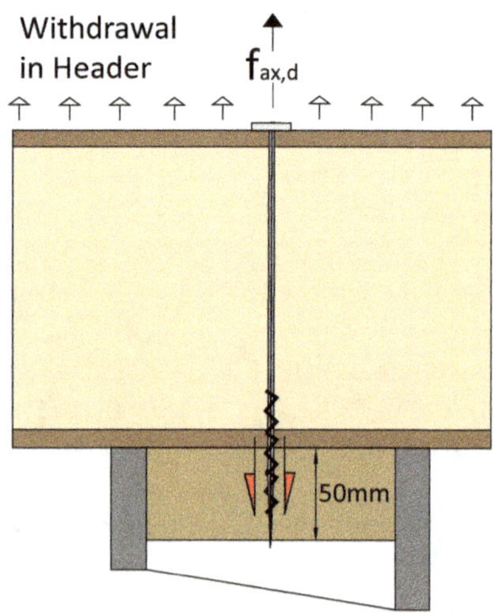

Figure 2.12.7 Withdrawal Failure Mode

For a 6mm screw, **figure 2.12.7**, penetrating 50mm into a C24 (50mm header) in a wall, the characteristic withdrawal resistance is:

$f_{ax,90,Rk} = 1.0x(0.52x6^{-0.5}x50^{-0.1}x350^{0.8}) \times 6x50x(6/8)$
$= 1.0 \times (15.57) \times 225 = 3503N$ (3.5kN)

Design resistance, instantaneous (C24) for service class 2:
$f_{ax,d} = f_{ax,90,Rk}.K_{mod}/\Upsilon_{m(t)} = 3.5 \times 1.1/1.3 = 2.96kN \approx$ **3.0kN** > 2.9 - *core bearing is critical*

Screw Fixing Uplift Failure Summary

From the three calculations for the screw fixing with 50mm screw 'pointside' penetration, the critical failure is *core bearing*. In some circumstances, depending on screw head diameter or 'pointside' penetration, withdrawal resistance could be less than core bearing resistance. All three checks should be undertaken. Uplift resistance of the panel may be increased by:

- Closer screw centres.
- Load spreaders/washers under the screws.

Eaves/Gable/End Panel with Timber Spline

A typical detail is shown in **figure 2.12.8,** which has a timber spline at the end of the panel with a fixing screw into the wall below. This is the most common condition for an end panel. The stud within the panel prevents distortion of the foam, for panel bearing or screw uplift.

The critical resistance is governed by the lesser of 'pull through' or withdrawal failure.

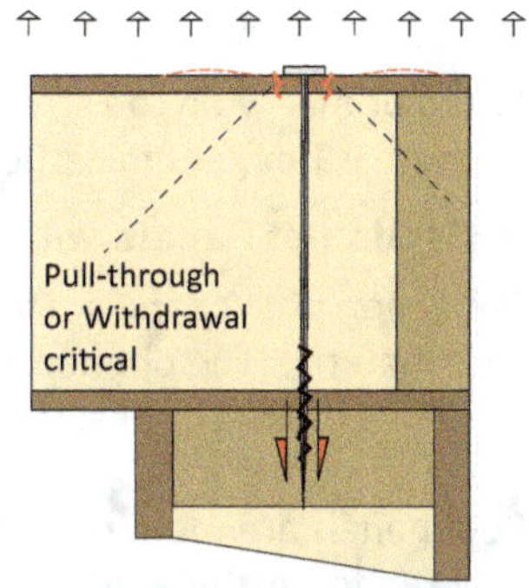

Figure 2.12.8
Typical Eaves Detail

2.13 OTHER COMMON SIP FACE MATERIALS

Steel Faced SIPs

Thin gauge steel faced SIP panels are used frequently in insulated walls in cold rooms, providing a clean, robust partition.

Properties: Face thickness 0.5 - 0.9mm
E_f = 210,000 N/mm², G_f = 80,000 N/mm², f_t = 275N/mm²
Creep - negligible (K_{def} = 0)

Aluminium Faced SIPs

Aluminium faced SIP panels have been used for floors, sometimes with box section or timber splines. In Africa, termites are a problem with OSB skins; aluminium is an option to resolve this.

Properties: Face thickness 0.9 - 1.7mm
E_f = 85,000 N/mm², G_f = 32,000 N/mm², f_t = 250N/mm²
Creep - negligible (K_{def} = 0)

Magnesium Oxide (MgO) Faced SIPs

Aluminium faced SIP panels are very useful for walls, particularly in Africa, where termites are a problem with OSB skins. Because MgO boards are brittle, and have relatively low tensile strength, they are generally not suitable for roof panels.

Properties: Face thickness 9 - 12mm
E_f = 4500 N/mm², G_f = 2200 N/mm², f_t = 3.5N/mm² (f_c = 11N/mm²)
Creep - K_{def} similar to OSB - *check with manufacturer.*

Chapter 2

Quick Note Summary

Bending Strength

Based primarily on the strength of the skins, core strength may be ignored. Bending strength is not generally the critical design condition.

Deflection of Panels with SIP Splines

Deflection of SIP panels is influenced strongly by the shear properties of the foam core material. The phenomenon of creep causes long term deflection in panels, which most often governs design.

Roof panels are more common than floor panels in SIPs. Floor panels should be checked for instantaneous deflections to ensure that the fundamental natural frequency is not too low and the floor is not too lively.

Deflection of Panels with Timber Splines

Timber splines in lieu of SIP splines significantly stiffen panels and provide robust fixing points. A drawback is that the timber is a cold bridge, raising the U value compared with panels with SIP splines.

Continuous Spans

Panel layouts can benefit from continuity, which can significantly reduce deflections. Laying panels in a chequerboard fashion with end joints at 1/4 to 1/3 span positions may be adopted to avoid moment connections.

2 Way Span Action

Panel assemblies can benefit from two way span action, which can reduce deflections for three or four sided support conditions. The secondary span may be modelled with 50% effective modulus based on the primary span direction. Orthotropic plate elements are useful in this respect in plate elements in finite element models.

Voids in Panels

Panels may be trimmed with timber or steel elements, as timber splines, or deeper if necessary.

Shear at Panel Supports

Where panels have embedded timber splines at supporting walls or beams, the shear strength may be derived from test data. If timber splines are not present at supports, then the core compression resistance is critical to the resistance of the panel at a support. This is not common in end panels but is a consideration for panels with SIP splines on continuous internal supports.

Local or Punching Shear

Core compression resistance is generally critical to locations of local bearing or point loading where timber splines are not present to strengthen the panel.

Other Facing Materials

Steel, Aluminium and Magnesium Oxide boards are common SIPs facing alternatives.

Chapter 3

Stability - Racking (In-Plane Shear)

Lateral stability. Racking in SIPs and sheathed stud walls. Standard methods of assessment. Examples: panels and house appraisal. Supplementary framing in timber, glulam, steel. Finite element models.

3.1 RACKING (IN-PLANE SHEAR)

Wind loads or earthquakes cause horizontally applied (or lateral) forces to act on the external walls of buildings. The capacity of the framing must be checked to ensure that the building is capable of resisting lateral loading. This is referred to as racking. In the UK, manufacturers publish the racking strength of panels based on standard testing. SIPs panels constructed with OSB faces have racking resistances comparable with double sheathed stud walls.

If a building has large openings, then these may be strengthened with steel or timber framing to substitute for, or complement the overall racking resistance. Internal sheathed timber frame stud walls may also assist.

3.1.1 Racking Analytical Method

The method followed in this book is based on British Standard BS5268 Part 6.1, with tabulated test values consistent with that code for sheathed stud walls, which is a permissible load design rather than a limit state. PD-6693-1, when evaluating the masonry cladding shielding effect, should be used in preference to BS5268 Part 6.1. Factors and coefficients for sliding and overturning are consistent with Eurocode 5 (EC5).

PD 6693-1 is advocated for platform timber frame walls in the UK. For SIPs BS5268 Part 6.1 accords with permissible racking strength test data provided by SIPs manufacturers and form of construction (see **section 3.2.1**). In load tests at Edinburgh Napier University (Dhonju[11]) EC5 method B (equivalent to BS5268 Part 6.1) demonstrated panel adequacy. PD-6693-1 and EC5 rules for panel opening ratios places limitations on walls. Porteous J. & Kermani A. (2013) Structural Timber Design to Eurocode 5 (see s**ection 1.7**) notes that PD-6693-1 is applicable if the walls are not anchored at their ends, as advised in the UK National Annex to EC5.

EC5, section 9.4.2, has two methods: Method A (walls anchored at the ends) and method B (based on BS5268 Part 6.1). Method B has inaccuracies. Furthermore, a paper by Le Kuai12 et al., from Linnaeus and Karlstad University, investigating sheathed walls, noted that EC5 is 'too conservative for short wall elements with openings'. BS5268 part 6.1 has been adopted herein for racking on the basis that it accords with SIP manufacturers' test methods, has been widely tested and is well established.

In the **section 3.2**, the method employed for stability checks is summarised. Following this are examples which relate to buildings no more than three stories with external SIP walls and internal SIP with stud walls are assessed where needed for stability.

3.1.2 Influence of Frame Construction (Platform or Partial Wall Continuity)

It is worth making a note with reference to racking document PD 6693-1. This method is largely recommended for *'platform frame construction'*. In the UK, this form of timber frame construction predominates. Successive floors sit on load bearing walls, providing a deck for the upper walls, which sit on soleplates supported off joists with timber blocking, or rim boards, as illustrated in **Figure 3.1.1 and 3.1.2** below. See also **figure 8.3.6**.

A large number of SIPs buildings in the UK have wall and floor separation, as described in **section 1.4**, which allows good shear interaction between upper and lower walls. This enables the designer to consider additional complimentary shear interaction where more than one storey contributes to stability.

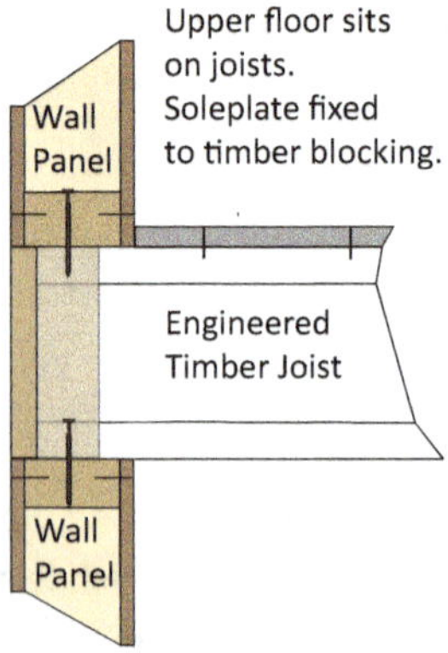

Figure 3.1.1 Platform Frame External Wall Section

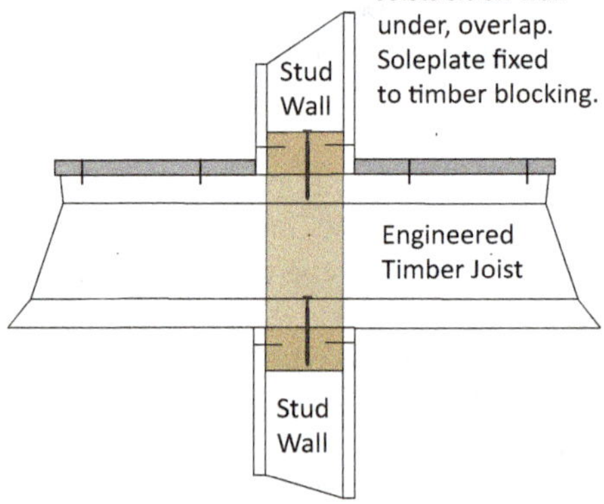

Figure 3.1.2 Platform Frame Internal Wall Section

Even when internal upper and lower walls are separated, as in **Figure 3.1.2**, external SIP walls are commonly constructed with upper and lower lift walls tied through the floor deck (see **Figure 1.4.6 and 1.4.10** in **chapter 1** and below). This has two benefits: improved insulation without joist-wall penetration and robust wall shear interaction with opportunities for wall posts to interconnect to tie down walls, wholly or partially, if required.

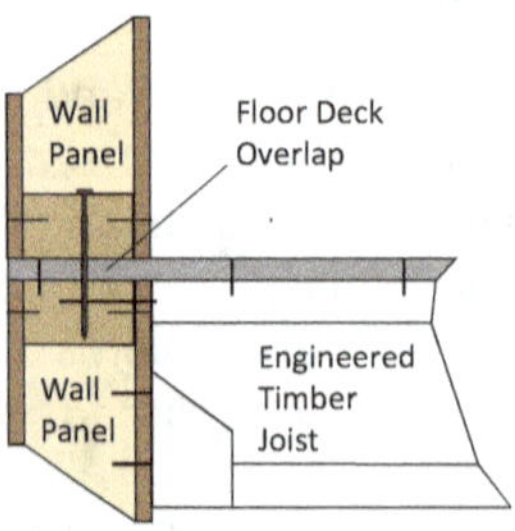

Figure 1.4.6 (from Chapter 1)

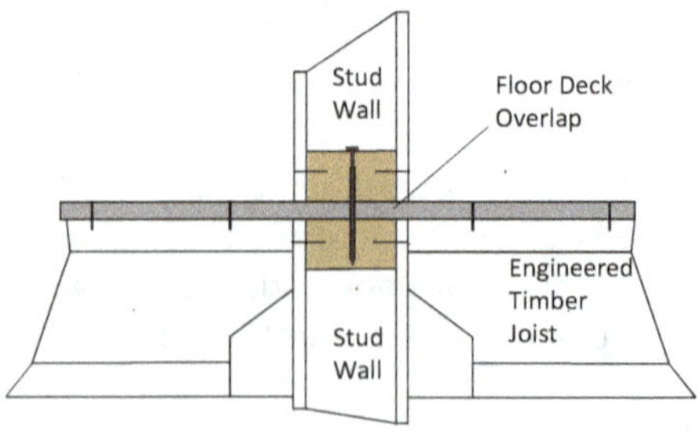

Figure 1.4.10 (from Chapter 1)

3.2 SUMMARY OF STABILITY ASSESSMENT TO BS 5268 - PART 6.1

BS5268 Part 6.1 gives a full outline of the methodology for checking the racking for sheathed walls. This has been adapted for SIP walls, whereby the shear capacity published by SIP manufacturers, based on testing, is generally based on tests in accordance with BS EN594:1996.

3.2.1 Design to Recommended Racking Capacity (Unfactored)

Throughout the book, except for racking, Eurocode 5 limit state design is employed. The reader should take careful note of the difference between ultimate resistances and recommended (or permissible) resistances. Racking resistances for sheathed stud, using this method, are obtained from Table 2 in BS 5268 Part 6.1, assuming nail spacing, sheathing and plasterboard as specified.

To obtain a conservative design from permissible resistances, a simple pro-rata relationship based on the *lowest ultimate partial load factor* of 1.35 is assumed. A less conservative approach would be to use a *leading variable live factor* of 1.5.

For instance, if the recommended resistance of a SIP wall is 5.4kN/m as quoted in the Structural Building Systems BBA certificate, then the design resistance is obtained as below. Rv is the unfactored, or recommended (or permissible), racking resistance, and Rdv is the design resistance::

R_{dv} (design) = 1.35 x R_v = 1.35 x 5.4 = 7.3kN/m

Typical Recommended Racking Resistances, Rv, for Reference, are:

Obtained from manufacturers BBA agrément certificates:

SBS SIPs panels (110mm - 200mm thick)	5.4kN/m
Kingspan TEK (142mm - 172mm thick)	4.3kN/m
SIPs ECO panels (EPS core)	2.5kN/m

Obtained from BS-5268 - Part 6.1 Table 2 (*fixings as specified*):

Stud walls, single sheathed with 9mm OSB	1.68kN/m
Stud walls sheathed: 9mm OSB, 12mm plasterboard other face	1.96kN/m
Stud walls with 12.5mm plasterboard both faces	1.35kN/m

3.2.2 Simplifying Design Procedure

Given that SIP walls are circa twice as stiff as internal single sheathed stud walls, many domestic buildings may be demonstrated to satisfy stability criteria by ignoring internal stud walls as a simplification. Professional engineers work to deadlines and, mostly, fixed fees, which are quoted before the design progresses, often with limited information. If the external walls of a building suffice for stability resistance, this can save time, with less concern for future internal changes. Designers must always bear in mind that this might be limited by the capacity of the *floor diaphragm* to transfer load in wide buildings. Floor diaphragm calculation checks are discussed in this chapter. See **section 3.3**.

3.2.3 Laterally Loaded Panel Summary of Criteria

When selecting walls for racking checks in order to minimise calculations the designer should make a careful assessment of walls, taking into account any voids in floors which may reduce connectivity. Often, only ground floor walls need to be checked, which will be covered in examples in this chapter. The less experienced designer, if in doubt, should verify the building as a whole until sufficient confidence is gained in the procedure. Having selected appropriate walls to examine, the following criteria must be satisfied:

Figs 3.2.1 and 3.2.2 illustrate a wall panel comprising two SIP wall panels. Standard test panels are configured with standard 1.22mm wide panels at a height of 2.4m. This forms a square assembly.

Panels are anchored to the base with a sole plate fixed to a masonry support wall or concrete plinth. See **figures 8.3.3** and **8.3.4** for typical sole plate holding down details.

Lateral or sliding resistance is provided by friction (coefficient of friction from TRADA[13], μ = 0.25) and shear capacity of the line of anchors fixing the soleplate.

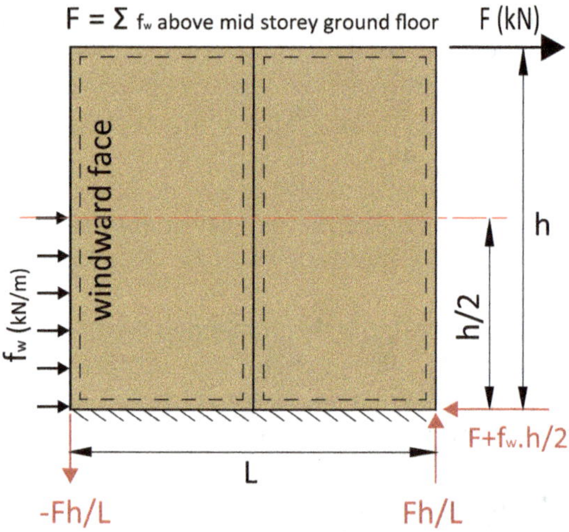

Figure 3.2.1 Wall Free Body Diagram Force F Applied

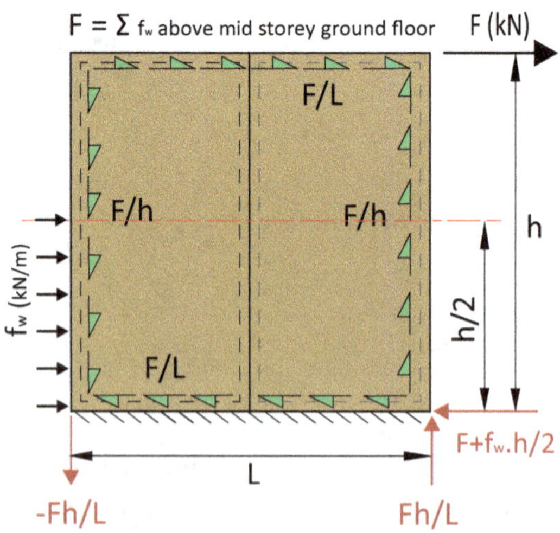

Figure 3.2.2 Wall Free Body Diagram Internal Shear Flow

Uplift at the left hand bottom corner is resisted by a combination of anchor, dead load (uniformly distributed, and any point loads).

Figure 3.2.1 shows a force F applied laterally. This induces an *overturning moment*, which is transferred within the perimeter fixings as a shear flow (see **Figure 3.2.2**) represented by green arrows. The overturning moment must be balanced by a *restoring moment*, which is provided by means of dead load (uniform or point load), and if necessary, the line of anchors at the base.

3.2.4 Design Procedure Summary to BS5268 Part 6.1

Racking verification of walls entails a three part procedure, to ensure that external loads balance and there is sufficient strength and stiffness in the panel to resist the loading:

Sliding Resistance (Section 3.2.5)

Calculate resistance from friction and base fixings.

Overturning Moment (Section 3.2.6)

Check restoring dead load/base fixing combination against overturning moment.

Racking Strength (Section 3.2.7)

Check panel resistance, taking into account openings, vertical loading and dimensional parameters.

3.2.5 Sliding Resistance Check

In **figure 3.2.3** below, the load denoted F(total) acts on a wall that is anchored to a supporting base - wall or slab, defined as:

F (total) = ∑ Lateral loads acting on that wall

F is the total of the lateral uniformly distributed load (UDL) over the top half of the lower wall and the full height of any walls and roof above, with the addition of the remaining wind UDL on the lower half in ***figure 3.2.3****.*

Resistance to sliding is provided by friction and anchor resistance, which is calculated based on the total dead load acting on the wall:

Dead Load, w = ∑ Wall, floor, roof & cladding self weight + Anchor resistance

When checking roof loading, this must be checked as a net load, which may be an uplift when subjected to wind loading - i.e., subtracted from the component loads.

If clad with masonry, the contribution may be calculated as a fraction of the total weight of the wall over its height. Standard tie spacing for timber frames is 450mm vertically and 600mm horizontally, which equates to 3.7 ties/m2. A safe approximation based on screw capacity in shear for instantaneous loading is a resistance of 25% of a masonry wall contributing to the total dead load. See **section 6.3.**

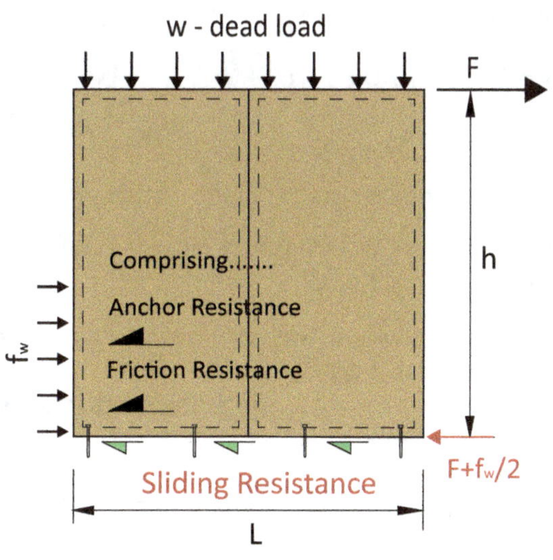

Figure 3.2.3 Wall Elevation Sliding Resistance

Calculation of Friction Resistance

From **section 3.2.3**, the friction coefficient may be taken as $\mu = 0.25$, with vertical dead load w:

Friction Resistance, $R_{frv} = \mu.w.L$

For example, if a 2.4m high wall, 2.4m long, weighs 1kN/m², and carries roof joists with a dead load of 0.80kN/m², spanning 4m (half loaded width acts on wall).

Dead load, $w = 2.4 \times 1 + (4/2) \times 0.80 = 4.0$kN/m

Friction Resistance, $R_{frv} = \mu.w.L = 0.25 \times 4.0 \times 2.4 = 2.4$kN

Calculation of Horizontal Anchor Resistance

If the wall is an upper storey, typically the header plate from the wall below is fixed to the sole plate of the wall above with long screws. Capacity of screws is outside the scope of this chapter, but may be obtained from Eurocode 5 tables or by calculation, noting that the design capacity is an ultimate condition. For example, 6mm screws 75mm long, connecting a 50mm deep soleplate to a 50mm header in the wall under have a design shear capacity of 1.5kN for an instantaneous load (wind). Based on **section 3.2.1**, using EC5 nomenclature for design capacity:

$F_{v,r}$ (permissible) = $F_{v,rd}$ (design) / 1.5 = 1.5/1 = 1.0kN per screw

Sliding resistance between of the panel from the fasteners (anchors), f*or a spacing, given by* S_f *is therefore*:

$$F_v = \frac{F_{v,r}}{S_f}.L$$

For example, if $F_{v,r} = 1.0$kN and spacing, $S_f = 350$mm (0.35m), then:

$F_v = 1.0 \times 2.4 / 0.35 = 6.9$kN

Combined Sliding Resistance

Total combined resistance of the wall example in this section is, therefore:

Total resistance to sliding $F_{sv} = R_{frv} + F_v = 2.4 + 6.9 = 9.3$kN

Verification of Static Equilibrium

To verify equilibrium, the more onerous Eurocode 5 factors are applied. Comparing the unfavourable condition of dead load (permanent) factor $\gamma_{gj,inf} = 1.0$, with live (variable) factor $\gamma_Q = 1.5$, for sliding resistance and applied load, respectively. This condition must be satisfied:

$0.9\ F_{sv} \geq 1.5\ F$ ...where F = lateral load acting on wall shown in **figure 3.2.3**.

In the example above, 90% of lateral force F must not exceed (9.3 x 1.5) = 14.0kN

Note if Adhesive is Applied at the Interface Between Walls

Shear capacity at the interface of soleplate and header of upper floors may be further enhanced by a site applied adhesive. Generally, adhesives are stronger than the constituent timber. Even with a 50% reduction, allowing for debonded areas gives a value for C16 grade timber of circa 0.5 N/mm² bond strength, which would give, for a 125mm wide soleplate a shear resistance of 62kN/m; well in excess of normal wind loads. A safe assumption, if site construction method is unknown, is to ignore this enhancement, but it is worth keeping in mind, in case special conditions need additional strength.

3.2.6 Overturning Check

Figure 3.2.4 illustrates a panel assembly with lateral load, denoted F_{OT}, with dead loads which comprise a uniformly distributed load (UDL) w and three point loads. Point load P1 acts in the centre of the panel and P2 at the right hand end. Force F_{OT} induces a moment which pivots about point B at the base.

Calculation of Restoring Moment (About point B)

Over-length, L, with P_1 applied at the centre, and P_2 at the ends:

Restoring Moment, $M_R = w.L^2/2 + P_1.(L) + P_2.(L/2) + P_1.(0)$

Calculation of Overturning Moment (About point B)

Over height, h, with applied overturning load at the top over the wall, F_{OT}:

Overturning Moment, $M_{OT} = F_{OT}.h$

Overturning Verification

If this overturning moment, with applied partial factor ($\gamma_Q = 1.5$), exceeds the sum of restoring moments, based on the dead load combination, with applied partial factor ($\gamma_{gj,inf} = 1.0$), then additional holding down measures will be needed; generally by soleplate fixings uplift resistance, and straps fixed to the base support, if anchors are insufficient.

$0.9\ M_R \geq 1.5\ M_{OT}$

An example follows overleaf.

Methods of strapping or holding down are shown in **figure 8.3.4**.

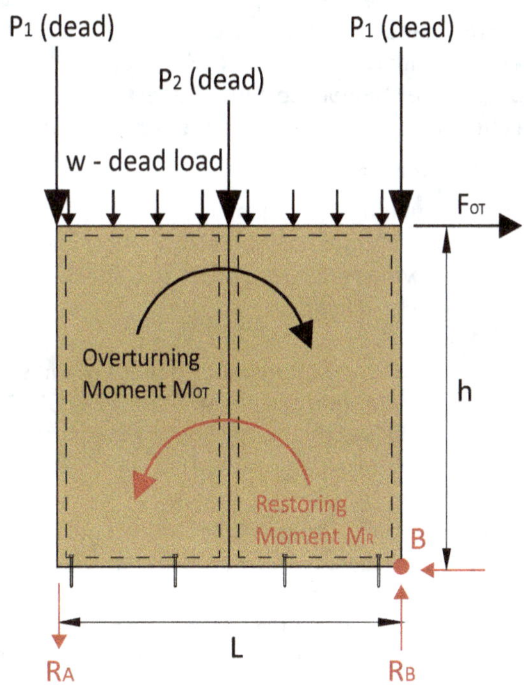

Figure 3.2.4 Wall Elevation Overturning Check

Overturning Check Example (Based on Figure 3.2.4)

A 2.4m x 2.4m wall with UDL, w = 5kN/m, no central point load (P_2 = 0) and end load, P_1 = 4kN and has a lateral force, F_{OT} = 6kN:

Overturning Moment, $M_{OT} = F_{OT}.h = 6 \times 2.4 = 14.4$kNm
Restoring Moment, $M_R = w.L^2/2 + P1.(L) + 0 = 5 \times 2.4^2/2 + 4 \times 2.4 = 24$kNm

Check Condition 0.9 $M_R \geq 1.5\ M_{OT}$,

$0.9M_R = 0.9 \times 24.0 = 21.6$kNm
$1.5M_{OT} = 1.5 \times 14.4 = 21.6$kNm (= M_R)

Wall just satisfies the overturning check.
Anchors at the base do not need checking for uplift resistance.

3.2.7 Racking Resistance

Racking resistance, as discussed in the introduction, is based on British Standard BS5268 Part 6.1. Racking capacity for SIP and sheathed stud walls may be obtained from **section 3.2.1** and table 2 in BS5268.

Wall Racking Capacity

Panel racking resistances are based on a standard wall panel assembly 2.4 x 2.4m, which is tested to verify strength. The capacity is based on the lesser of lateral stiffness and racking strength (divided by a safety factor) when a test wall is subjected to repeated lateral loading. The quoted permissible racking load resulting from tests on a SBS panel is 5.4kN/m, from **section 3.2.1**, noting that the actual failure load (with a standard safety factor of 2.4 omitted) was circa 30kN/m. For a 5.4kN/m load applied, the standard test wall deflected at the top at approximately 5mm, which is useful for calibrating finite element plate models, as can be seen in **section 3.6**.

Modification Factors

When calculating the racking capacity of a wall, the basic racking capacity is factored to take into account its geometry, vertical loading, nail spacing (not applicable to SIP panels which are constructed to a specification consistent with test walls), and a permissible load factor:

K101 and **K102**: Variation in nail diameter and spacing, respectively, from standard 3mm nails at 150mm spacing in Table 2 of BS5268 Part 6.1. For the purposes of this book 3mm nails with 150mm spacing is assumed; hence K101 =1.0 and K102 = 1.0 and may be ignored.

K103: Variation in sheathing board thickness. As noted for K102, this is assumed to accord with Table 2 (K103 = 1.0), and may be omitted from checks.

K104: Modification for wall height ratio. For wall panels less than 2.1m and greater than 2.7m in height, this should not be extrapolated.

$$K104 = \frac{2.4}{h}$$

For walls greater than 2.7m high, finite element models for external SIP walls have been analysed, whereby consideration of internal shear flows have been checked against design details. **See section 3.6**.

K105: Length Modification Factor

For walls less than 2.4m long

$$K105 = \frac{L}{2.4}$$

For walls between 2.4m and 4.8m long

$$K105 = \left(\frac{L}{2.4}\right)^{0.4}$$

For walls greater than 4.8m long

$$K105 = 1.32$$

K106: Allowance for window and door openings, whereby the ratio of opening area to total panel area should not exceed 0.75 (openings are limited to 75% of wall area). All openings should be framed by studs. *Over deep openings, such as doors, there has to be a means of transferring horizontal shear, and the net vertical component of complimentary shear flow should be checked against lintel to support stud fixing.*

$K106 = (1\text{-}1.3\rho)^2$ .....where $\rho = A_a/A_t$

A_a is the total area of openings in the wall
A_t is the total area of the wall, including openings

K107: Racking capacity enhancement from vertical load. This can be expressed as an empirical formula:

$K107 = 1+(0.09.w + 0.0015.w^2).(2.4/L)^{0.4}$

where w = UDL vertical load, L = wall length

Intermediate Point Load Equivalent UDL

Where there is an intermediate point load (see P_1 in **figure 3.2.5**), it is possible to convert this to an equivalent UDL. From **figure 3.2.5** for intermediate load, distance A from the leeward face:

$W_{eq} = 2.a.P_1/L^2$

K108: Factor of 1.10, which pertains to the stiffening effect of internal walls.

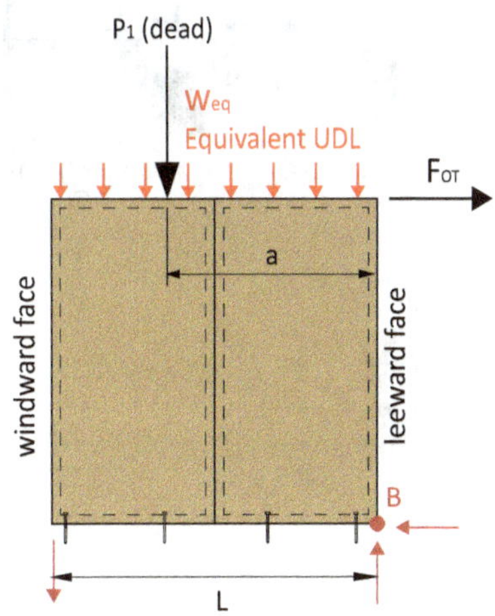

Figure 3.2.5 Equivalent UDL

Calculated Racking Resistance of a Wall

As noted in the previous pages, modification factors K101-K103 may be assumed to be unity for SIP walls, and stud walls complying with table 2 of BS 5268 Part 6.1. The overall racking resistance of a wall may be calculated from:

Racking Resistance, $R_{rk} = R_v \times L \times K_m \times K_w$

Where...

R_v is the basic racking resistance (see **section 3.2.1** summary)
L is the wall length
K_m is the product of modification factors K101, K102 and K103 (1.0)
K_w is the product of modification factors K104, K105, K106, K107 and K108

For example, in **figure 3.2.6**, the wall is 4.8m long with two 0.80 x 1.0m openings, 2.4m high, with R_v = 5.4kN/m, vertical loading, w = 2kN/m, then:

K104 = 1.0 (2.4m high)
K105 = $(4.8/2.4)^{0.4}$ = 1.32
K106 = $(1-1.3\rho)^2 = (1-1.3\times0.14)^2$ = 0.67 {where ρ = (2x0.8x1.0)/(2.4x4.8) = 0.14}
K107 = $1 + (0.09\times2 + 0.0015\times2^2).(2.4/4.8)^{0.4}$ = 1 + 0.19 x 0.76 = 1.14
K108 = 1.10

Racking Resistance, $R_{rk} = R_v \times L \times K_m \times K_w$

where K_m = 1.0 and K_w = K104xK105xK106xK107xK108

$R_{rk} = R_v \times L \times K_m \times K_w$ = 5.4 x 4.8 x 1.0 x (1.0 x 1.32 x 0.67 x 1.14 x 1.10) = 28.7kN

Applied load, F = 5.5kN < 28.7...*F is unfactored as this is a permissible load check. The wall has sufficient racking resistance.*

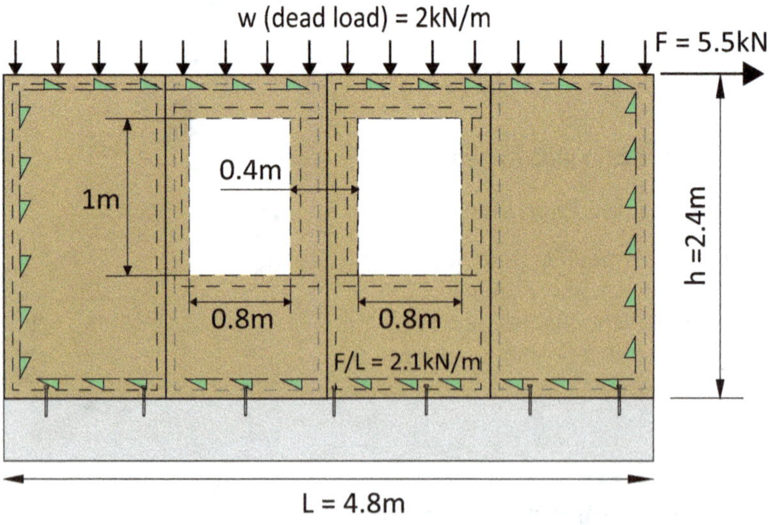

Figure 3.2.6 Wall Elevation Racking Example

3.2.8 Racking Wall Limitations to BS 5268 - Part 6.1

Openings within walls, under certain conditions, have limitations. The figures below summarise conditions in which panel racking parameters should be modified.

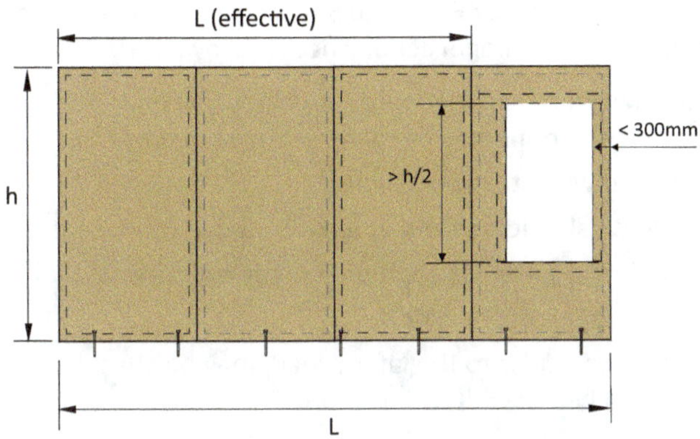

Figure 3.2.7 Wall Elevation Hole Close to Edge

In **figure 3.2.7,** a hole is located closer than 300mm from the edge of the panel. The panel should be treated as having an effective length up to the internal vertical edge of the hole - if the height of the hole is greater than half the height of the panel.

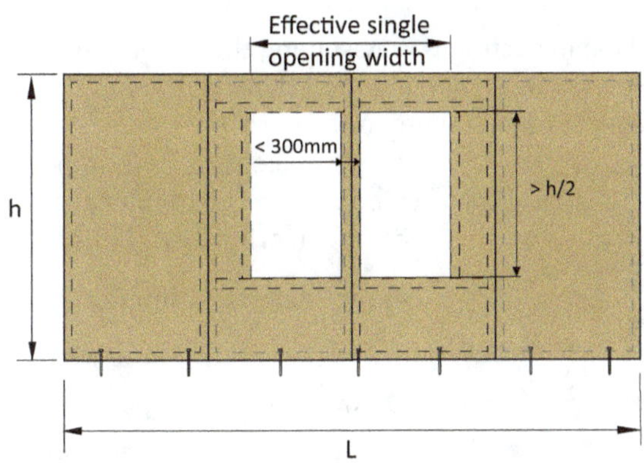

Figure 3.2.8 Wall Elevation Adjacent Holes

In **Figure 3.2.8,** two adjacent holes are located such that adjacent vertical edges are less than 300mm from each other. If the height of the hole is greater than half the height of the panel, then the panel is treated as having a single hole with width as the distance between outer edges.

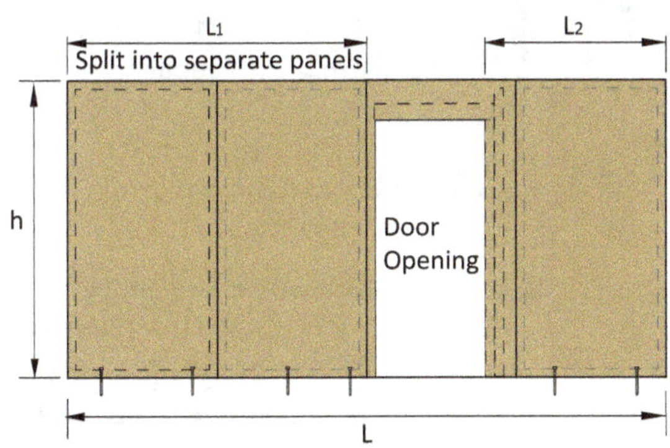

Figure 3.2.9 Wall Elevation With Internal Door

In **Figure 3.2.9,** an internal door limits transfer of complimentary shear at the head of the door. *If a strengthening frame is not adopted, or there is no solid sheathed wall in an upper storey over the opening*, the panel should be split into two separate panels with lengths, L_1 and L_2, as indicated.

3.3 FLOOR DIAPHRAGM ACTION

Suspended floors above ground floor level must be able to transfer lateral loads to the stabilising walls. Both BS5268 Part 6.1 and Eurocode 5 (section 9.2.3) refer to floor diaphragms. The reader is encouraged to check Eurocode 5, which gives guidance to simplified analysis. A brief summary of diaphragm action checks follows:

- Floor decks are commonly formed from particleboard (or chipboard) nailed or screwed to supporting joists. Joisted roof decks are generally plywood or OSB.
- Floor decks should be constructed with staggered panels.
- Fixings - improved nails or screws - should be fixed at a maximum 150mm centres
- The deck may be adhesive bonded to joists for additional shear resistance.
- Diaphragm action is analogous to deep beam action.
- Windward trimming beams, perpendicular to the lateral load, may be checked as 'compression chords' and leeward beams as 'tension chords'.
- Shear transfer to joists may be checked assuming, say 1m width, adjacent to the connecting shear wall.
- Around stair or service voids built-in trimming beams can transfer shear, but should be checked for reduced shear interaction if connected to a stability wall.

The house below is examined for racking in **section 3.4**. Assuming the outer external wall, noted in **Figure 3.4.6**, is 8m long and has an applied total shear at the first floor along its length of 12kN from wind loading. 60mm long improved 3.4mm nails at 150mm centres fix 22mm chipboard to the top flanges of the engineering joists under.

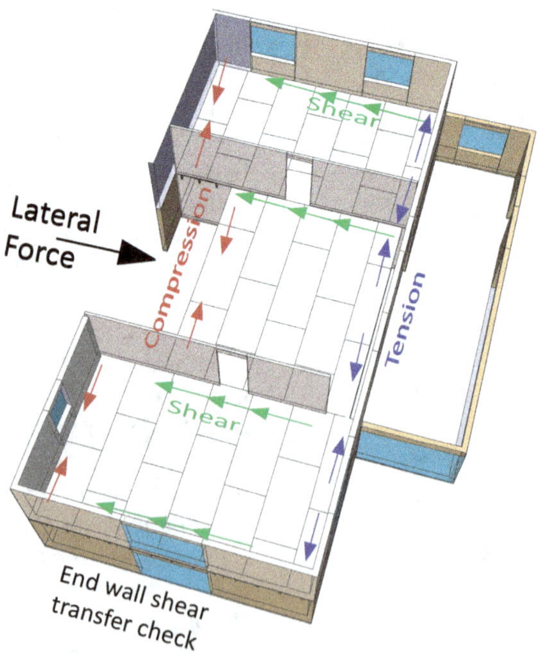

Figure 3.3.1 House First Floor Diaphragm

Shear Transfer Check:

Design shear flow in the wall
= 12kNx1.5/8m = 2.25kN/m

Instantaneous design resistance of single nail from EC5 ≈ 0.8kN

In 1m width adjacent to wall nail capacity = 0.80 x 1000/150 = 5.3kN/m > 2.25 applied.

Design shear resistance of 22mm chipboard assumed from EC5 is circa 1.1 N/mm^2.

Therefore shear resistance
= 1.1 x 22 = 24.2 N/mm (kN/m)
>> 2.25 applied. Ok.

Note floor overlaps walls, and joist fixings transfer shear into the wall locally. The floor diaphragm is sufficient.

3.4 TWO STOREY HOUSE RACKING ASSESSMENT

An example of racking assessment for a house follows. This is a house in a rural exposed location with reasonably high wind loading. External SIP walls are clad with a render board. There is a ceiling on first floor (FF) and at first floor a dead load which includes an under floor heating system.

Summary of parameters:

- Live loads are ignored.
- Only load bearing walls are shown for the racking assessment.
- Dead Load first floor = 0.65kN/m²
- Dead Load FF ceiling = 0.50kN/m²
- External Wall load = 1.00kN/m²
- Internal Wall load = 0.50kN/m²
- 140mm Dense Block sits under all load bearing walls.
- Resin anchors fix sole plates to sleeper (supporting) walls under.
- Joists span (North - South) in direction indicated on **Figure 3.4.3.**

Figure 3.4.1 South View of House

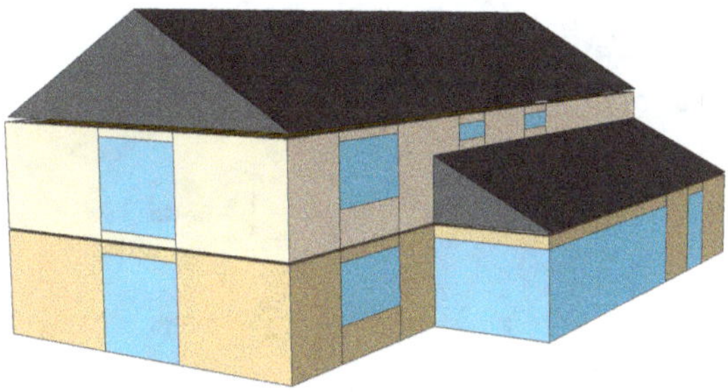

Figure 3.4.2 North View of House

Wall Construction:

External walls are 150mm SBS panels with a racking capacity of 5.4kN/m.

Internal walls are single sheathed with 9mm OSB sheathing in compliance with Table 2 of BS 5268 Part 6.1 with plasterboard facing to give a combined racking capacity of 1.96kN/m (1.68 + 0.28).

Wind Loading Assumptions:

- Roof wind uplift = Roof dead load. (*i.e. contribution ignored*)
- Roof lateral wind load contribution is insignificant.
- Wind loading in all directions is an overall value 1.10kN/m².
- Wind loading acts on the windward face and is uniform throughout the height.
- Wind acting in a North-South direction on the long side is not critical (i.e. sufficient external SIP walls are able to resist the load.
- Wind acting in an East-West direction is resisted by external gable walls and internal load bearing walls.
- Internal walls, as single sheathed, are 50% or half the stiffness of external SIP walls which are faced with OSB on both sides.

3.4.1 Loading on Stability Walls Resisting Wind Load (East/West Direction)

A dimensioned plan of the ground floor (GF) is shown below for the walls resisting a net wind load of 1.1 kN/m² in **Figure 3.4.4**. Joists span perpendicular to the walls so that their combined dead load from the floor assist in the racking resistance. See **Figure 3.4.3**.

Figure 3.4.3 House Joist Orientation

Wall Load Distribution

Since the building has an almost symmetrical orientation of shear walls, the wind load may be assumed to act in the centre.

A more rigorous assessment for asymmetrical plans involves calculating the centre of wind action based on the calculated centre of loading analogous to beam section properties. See **section 3.4.3**. Engineers are encouraged to review projects in a more rigorous manner to gain familiarity before making simplifications.

South Wall Idealised as 6m Solid Wall

The walls selected below are split between openings, aside from the external wall noted on the south side. This wall is 11m long (by 2.6m high) with 3 openings totalling 4.7m². Based on a ρ = 0.16 (i.e. 4.7/11 x 2.6), and K106 value (see **section 3.2.7**) of 0.63 the equivalent length, assumed as *approximately* 6m solid in calculations.

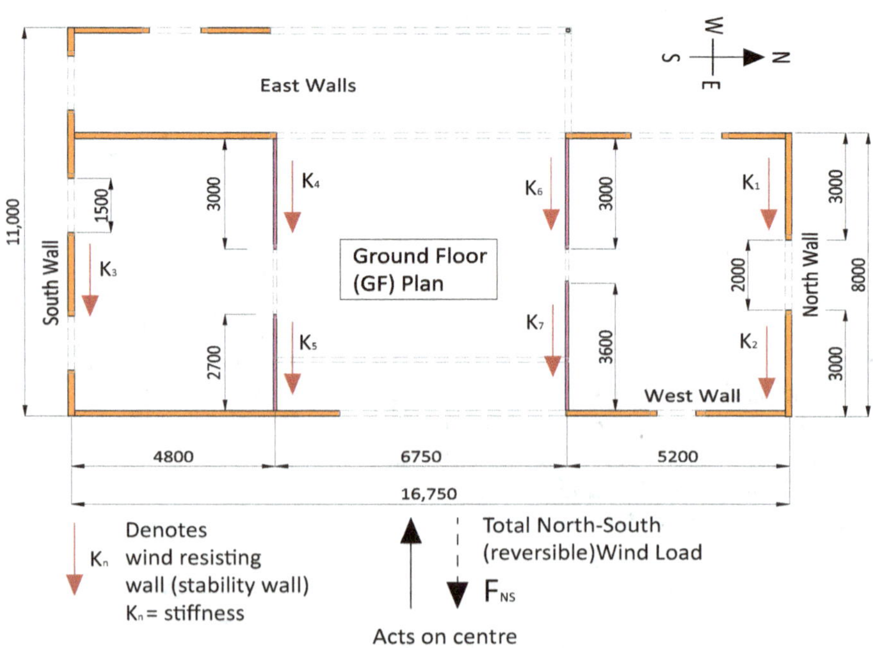

Figure 3.4.4 House Ground Floor (GF) Plan

Apportioning Loads to Individual Stability Walls

Having selected stability walls, the loading for each wall may be apportioned by calculating the equivalent stiffness of each wall on the basis of the following:

Stiffness of Each Wall, K_n, where...

Wall stiffness K_n *is proportional to wall length*:
$K_n \alpha L_{wall}$ (or K_n = Constant x L_{wall}).
This is an approximation based on the fact that shear stiffness predominates over bending stiffness.

Load resisted by stability wall, $R_{wall,n}$ is ratio of K_n to the sum of wall stiffnesses:

$$R_{wall,n} = K_n/\sum(K_1 - K_n)$$

For the seven stability walls, in proportion to their equivalent lengths in **figure 3.4.4**, **table 3.4.1** below gives a summary of calculated loading based on the total north-south wind load, P_{NS}, expressed as a vertical line load.

Where F_{NS} = wind load x Width of windward face = 1.1 x 16.75 = 18.4kN/m.

External SIP Wall Stiffness Identity	**Relative Stiffness Value Assuming K_4 = 100** K_n = Constant x L_{wall} Note $K_{SIP} = 2K_{stud}$	**Ratio of Wall Relative Stiffness (K)** $K_n / \sum(K_1\text{-}K_7)$	**F_{NS} = 18.4kN/m** *(=1.1 x 16.75)* **Load on Each Wall (Pro-Rata)** kN/m
K_1	200	0.17	3.1
K_2	200	0.17	3.1
K_3 (Equivalent 6m wall)	400	0.33	6.1
Internal Stud Wall Stiffness Identity			
K_4	100	0.08	1.5
K_5	90	0.07	1.3
K_6	100	0.08	1.5
K_7	120	0.10	1.8
Check	$\sum(K_1\text{-}K_7)$ Total: 1210	Sum: (1.0) *Ok*	Sum: (18.4) *Ok*

Table 3.4.1 Loading on Individual Walls by Relative Stiffness

Having found the individual wall loads, the two critical walls, based on length, to check are K_1 (same as K_2) on the east side and K_5 internal wall, which is the minimum length of the internal walls. If these have sufficient racking resistance, then it follows that the remainder of the walls will be adequate, which may be readily verified.

3.4.2 External Walls K_1 and K_2 Racking Check

Using the methods outlined in **section 3.2.5** to **3.2.7** for sliding, overturning and racking resistance, the walls will be checked. The opening between is carried by lintels at two levels. Lintel loading from the joists is beneficial to holding down the windward edge of each panel. **Figure 3.4.6** below illustrates wind and dead loading on both panels.

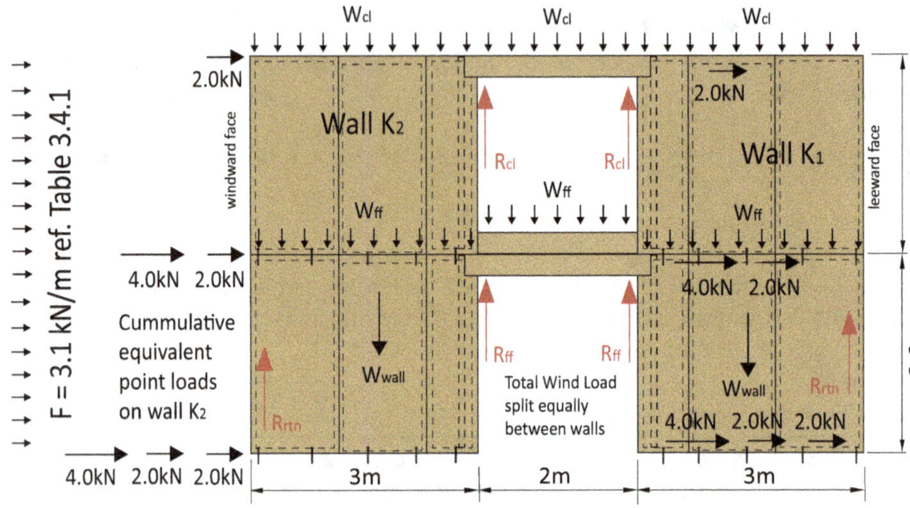

Figure 3.4.6 North Elevation - Loading on Walls K_1 and K_2

Derivation of Panel Loads

In the beginning of **section 3.4** overall dead loads are summarised. The floor span from **figure 3.4.4** is 5.2m, and wall height is 5.2m. Both walls have the same geometry and loading.

Ceiling load at eaves level: $W_{cl} = 0.50 \times 5.2/2 = 1.3$kN/m

First floor load: $W_{ff} = 0.65 \times 5.2/2 = 1.7$kN/m

Wall Load (2 Stories):

$W_{wall} = 1.0 \times 5.2 = 5.2$kN/m

Wind Load (from **Table 3.4.1**)

F = 3.1kN/m

Wind load is split into point loads at the ceiling, first floor and base levels. Each wall resists half the total wind load. The load accumulates from top to base.

F_{cl} (each wall) = $3.1/2 \times 2.6/2 = 2$kN

$F_{ff} = 3.1/2 \times 2.6 = 4$kN

$F_b = 3.1/2 \times 2.6/2 = 2$kN

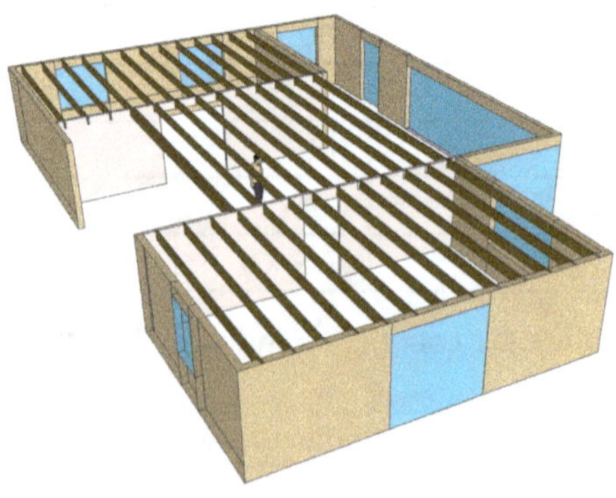

Figure 3.4.7 View on Ground Floor East

Racking (In-Plane Shear)

Evaluation of Point Loads at Wall Ends

The lintel reactions at theceiling and first floor level, R_{cl} and R_{ff} respectively, and the shear reaction from the wall fixings to the return external walls (on north and south faces), R_{rtn} may be included for the overturning checks. Since the wind load is reversible, the minimum is assumed. *By inspection of wind loading, the ground floor wall is critical.*

Lintel reaction at ceiling level, $R_{cl} = W_{cl}$.Lintel span/2 = 1.3x2/2 = 1.3kN
Lintel reaction at first floor level, $R_{ff} = W_{ff}$.Lintel span/2 = 1.7x2/2 = 1.7kN

Total reaction from lintels = $R_{cl} + R_{ff}$ = 1.3 + 1.7 = 3kN

Reaction from return wall (assume 1m width, w_d, full height), $R_{rtn} = W_{wall}.h.w_d$
R_{rtn} = 1.0x5.2x1.0 = 5.2kN (> 3kN)

For the analysis the lower reaction, based on $R_{cl} + R_{ff}$ will be adopted.

Sliding Resistance Check

From **section 3.2.5** resistance to sliding is calculated:

Total vertical load, W_T
= 1.3 + 1.7 + 5.2
= 8.2kN/m

Frictional Resistance
$R_{frv} = 0.9.\ \mu.\ W_T.L$
= 0.9 x 0.25 x 8.2 x 3
= 5.5kN

1.5 x Applied Load
= 1.5 x (4+2+2) = 12kN

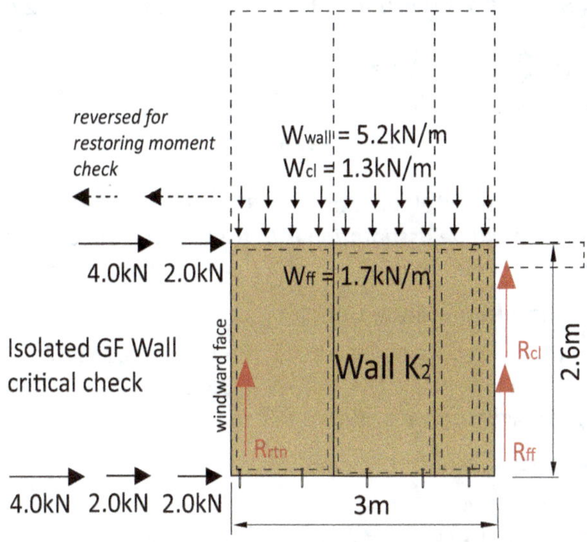

Figure 3.4.8 East Elevation - Loading on Wall K_2

Therefore anchor resistance required (*fixed to 7N/mm² solid block*) = 12kN - 5.5kN = 6.5kN (equivalent to 6.5/3 ≈ 2.2kN/m).

Assuming resin anchors are at 600mm (0.60m) centres, the shear resistance (unfactored) required, F_{vrd} = 2.2 x 0.6 = 1.3kN. *This is achievable with drilled concrete anchors, expanding bolts or resin anchors at centres shown in **figure 8.3.3**.*

Overturning Check (ref. Section 3.2.6)

Restoring Moment, *based on wind in opposite direction,* since $R_{rtn} > R_{cl} + R_{ff}$):
$0.9\ M_R = (W_{cl} + W_{ff} + W_{wall}).L^2 / 2 + (R_{cl} + R_{ff}).L = 0.9 \times 8.2 \times 3^2 / 2 + 3 \times 3 = 41.3$kNm

1.5 x Overturning Moment (at FF), $1.5.M_{OT} = 1.5 \times (4 + 2) \times 2.6 = 23.4$kNm < 41.3
No additional sole plate strapping needed for sliding or overturning.

Racking Resistance Check of Wall K2 (ref. Section 3.2.7)

Summary of factors K102 to K108 follows:

K102 = K103 = 1.0
K104 = 2.4/h = 2.4/2.6 = 0.92 (*height factor*)
K105 = $(L/2.4)^{0.4}$ = $(3.0/2.4)^{0.4}$ = 1.09 (*length factor*)
K106, no openings are present in the panel, K106 = 1.0
K107 = K107 = 1 + $(0.09.w + 0.0015.w^2).(2.4/L)^{0.4}$
Wall self weight omitted as this is based on loading onto the wall (W_{cl} + W_{ff} = 3kN/m)
K107 = 1 + $(0.09x3 + 0.0015x3^2)$ x $(2.4/3.0)^{0.40}$ = 1 + 0.28 x 0.91 = 1.25
K108 = 1.1

Basic racking resistance of SBS/Hemsec panel, R_v = 5.4kN/m

Racking resistance of panel, R_{rk} = R_v x L x K_m x K_w ...where k_m = 1.0 from **section 3.2.7**
= 5.4 x 3.0 x 1.0 x (0.92 x 1.09 x 1.0 x 1.25 x 1.10) = 22.3kN > 6kN (= 4 + 2)

3.4.3 Internal Wall K5 Racking Check

Calculations based on the evaluation of wall K2 in **section 3.4.2** for the internal stud wall gave the following results:

0.8m span Lintel reaction R_{cl} + R_{ff} = (6.75+4.8)/2 x (0.5+0.65) x 0.8/2 = 2.7kN
A conservative assumption is to assume a reduced first floor load of 1.6kN/m. The trimmer point load, Pt, and excess floor load to the right is ignored.

Sliding Resistance Check

Frictional resistance to sliding, R_{frv} = 0.9 x 0.25 x (2.6 + 2.9 + 1.6) x 2.7 = 4.3kN
< 1.5 x (1.7 + 0.8 x 2) = 4.95kN (> 3.4 anchors required to increase resistance)
Note: wall self weight above FF ignored

Overturning Check

Restoring Moment, M_R =
(2.6+2.9+1.6)x2.7^2/2+2.5x2.7
= 33kNm (0.9M_R = 29.7kNm)

1.5 x Overturning Moment,
M_{OT} = 1.5 x (1.7+0.8) x 2.6
= 9.8kNm < 29.7

Racking Resistance, R_{rk}
Where R_v = 1.96kN/m
from **section 3.4**....

R_{rk} = 1.96 x 2.7 x 1.5*
= 7.9kN > 2.5kN (=1.7+0.8)

**K104 = 0.92, K105 = 1.05, K107 = 1.41, K108 = 1.1, all other factors = 1.0*

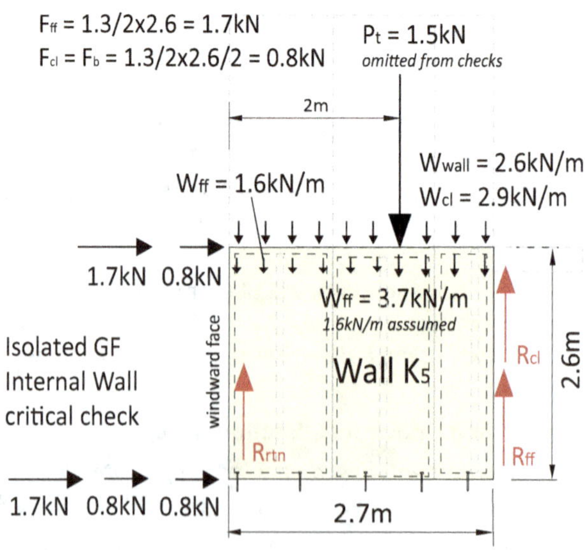

Figure 3.4.9 East Elevation of Wall K5

3.4.3 Asymmetric Plan With Eccentric Wind Load

In **section 3.4.1** the house plan was assumed to be symmetrical, which was a good assumption based on the distribution of selected stability walls. In **figure 3.4.10** and **figure 3.4.11** below, the 'L-shaped' floor plan, with walls which vary in length and, therefore stiffness, has a centroid which is offset from the centre of the layout.

Calculation of Centroid Position

In a similar way to beam cross section assessment, the centroid is calculated by the sum of wall moments about a chosen location. In the 'X' direction for the walls below the centroid position is calculated from the right hand wall (R_1) as the origin.

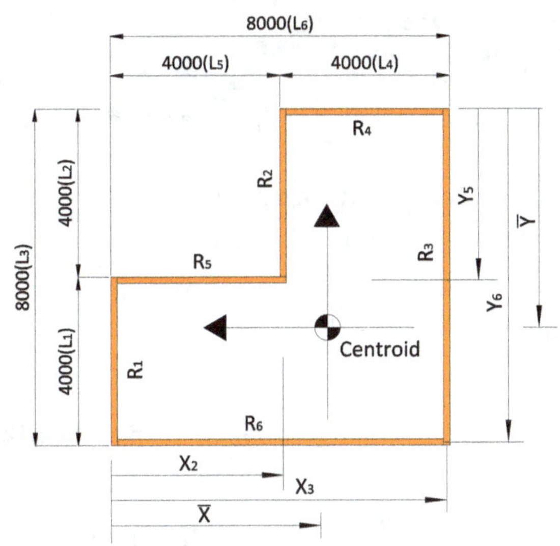

Figure 3.4.10 Asymmetric Floor Plan

$$\overline{X} = \frac{\sum(R_n).X_n}{\sum(R_n)}$$

As the stiffness may be assumed to be proportional to wall length, then the length of each wall may be substituted into the equation.

Assumption: $R_n = L_n$

It follows that:

$$\overline{X} = \frac{L_1.X_1+L_2.X_2+L_3.X_3}{L_1+L_2+L_3}$$

$$\overline{X} = (0x4 + 4x4 + 8x8)/(4+4+8)$$
$$= 5m$$

$\overline{Y}$ *derived as above.*

Calculating Eccentric Moment (Torsion)

If a wind load of 20kN acts on the centre of the layout, the action on the distribution of loading causes a clockwise moment (torsion) which is resisted by the anticlockwise reaction of all of the walls, as shown in **figure 3.4.11**.

The centroid for walls perpendicular to the wind (denoted in the Y direction), based on the geometry, can be calculated to be 5m from the top in **figure 3.4.10**.

Eccentricity (**figure 3.4.11**), $e = \overline{X}_{centre} - X = 5 - 4 = 1m$

Eccentric Moment, $M_e = F.e = 20x1 = 20kNm$

The reaction in each wall (R_{ne}) may now be determined.

Resultant Wall Reactions from Applied Force and Eccentric Moment

The reaction in each wall can be found by superposition.

Resultant reaction in each wall, $R_{nR} = F_{ne} + F_{nF}$ ..*see below for* F_{ne} *and* F_{nF}.

Eccentric Reactions in Walls (from figure 3.4.11)

Total eccentric moment is resisted by all racking walls and is proportionate to their stiffness (equivalent to length) and square of the perpendicular distance (X or Y) from the centroid.

If $F_{v,d}$ is the applied force, $F_{v,d}.e = K.\sum R_n.Z_n^2 \equiv K.\sum L_n.Z_n^2$

K is a constant, determined from the above
Z_n is the perpendicular distance (X or Y) from the centroid.
L_n is the wall length (assumed proportional to stiffness)

The force in each wall resisting the moment, $F_{ne} = K.L_n.Z_n$..*(note* F_{ne} *can be -ve)*

Applied Force Reactions Assuming no Eccentricity (from figure 3.4.12)

The reactions from applied force assume a centrally applied load distributed according to stiffness.

The force in each wall (assuming e=0), $F_{nF} = \dfrac{F_{v,d}.L_n}{\sum L_n}$

Walls perpendicular to the load do not contribute to the resistance, see **figure 3.4.12**.

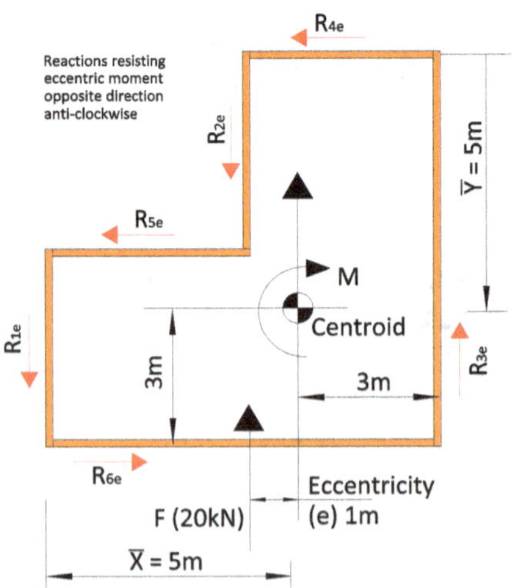

Figure 3.4.11 Eccentric Moment Reactions

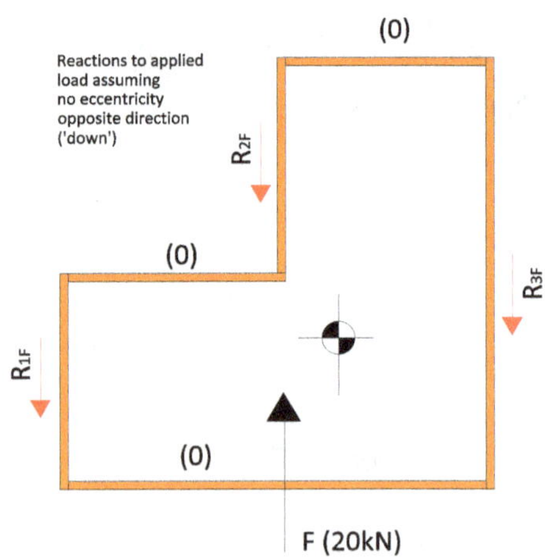

Figure 3.4.12 Applied Force Reactions

Resultant Racking Loads (Reactions) from Figure 3.4.11 and 3.4.12

Applying the formulae from the previous section, resultant reactions or racking loads in each wall may be calculated.

Find Eccentric Factor, K (working from bottom anticlockwise)

$K = F.e / \sum L_n.Z_n^2 = 20x1/(L_1.5^2 + L_2.1^2 + L_3.3^2 + L_4.5^2 + L_5.1^2 + L_6.3^2)$
$= 20x1/(4x5^2 + 4x1^2 + 8x3^2 + 4x5^2 + 4x1^2 + 8x3^2) = 20/352$
$= 0.057$

Racking Load in Each Wall from Eccentric Moment (Torsion)

Racking Load, $R_{1e} = R_{4e} = K.L_n.Z_n = 0.057x4x5 = 1.1kN$
Racking Load, $R_{2e} = R_{5e} = K.L_n.Z_n = 0.057x4x1 = 0.2kN$
Racking Load, $R_{3e} = R_{6e} = K.L_n.Z_n = 0.057x8x3 = 1.4kN$

Check: Total moment resistance = 2 (1.1x5m + 0.2x1m + 1.4x3m) = 20kNm

Note that R_{1e} and R_{2e} act in the same direction as R_{1F} and R_{2F} (i.e. +ve)
R_{3e} acts in the opposite direction to R_{3F} (i.e. -ve)

Racking Load Component Assuming no Eccentricity (e = 0)

Summation of wall lengths, $\sum L_n = R_{1F} + R_{2F} + R_{3F}$, *where R_{4F}, R_{5F} and $R_{6F} = 0$*

$\sum L_n = 4 + 4 + 8 = 16m$

$R_{1F} = F_{v,d}.L_1/\sum L_n = 20x4/16 = 5kN$
$R_{2F} = F_{v,d}.L_2/\sum L_n = 20x4/16 = 5kN$
$R_{3F} = F_{v,d}.L_3/\sum L_n = 20x8/16 = 10kN$

Resultant Racking Loads in Each Wall

$R_{1R} = R_{1e} + R_{1F} = 1.1 + 5 = 6.1kN$
$R_{2R} = R_{2e} + R_{2F} = 0.2 + 5 = 5.2kN$
$R_{3R} = R_{3e} + R_{3F} = 10 - 1.4 = 8.6kN$
$R_{4R} = R_{4e} + R_{4F} = 1.1 + 0 = 1.1kN$
$R_{5R} = R_{5e} + R_{5F} = 0.2 + 0 = 0.2kN$
$R_{6R} = R_{6e} + R_{6F} = 1.4 + 0 = 1.4kN$

Summation reactions parallel to wind,
R1R + R2R + R3R = 6.1 + 5.2 + 8.6 = 19.9kN ≈ 20kN ok

In conclusion, the eccentricity increases loading on the parallel walls furthest from the centroid (R_1 and R_2), and reduces loading on the long wall (R_3). Walls perpendicular to the applied wind load have negligible reactions.

Simplifying Analysis for Asymmetric Plans

The previous section involves complex assessments of wall contributions. External walls may all have openings. **Figure 3.4.13** below shows a condition where the long wall R_3 has 3 openings. This reduces the stiffness of the wall.

It is very often possible to simplify the analysis if the walls are capable of resisting higher racking loads In the example below, for simplicity, it is assumed that wall R_1 and R_3 resist half the racking load and the contribution from R_2 is ignored. The worst case is wall R_1. *Eccentricity is ignored*. If this wall is found to be adequate, then the conservative calculation is sufficient for purpose.

Simplified wall reactions, $R_{1F} + R_{3F} = 20kN$ & $R_{1F} \approx R_{3F}$
Since eccentricity is ignored, assuming e=0, *then* $R_{1R} = R_{1F}$ & $R_{3R} = R_{3F}$
Therefore, R_{1R} ($\approx R_{3R}$) = 20/2 = 10kN

From the previous page, R_{1R} = 6.1kN (< 10) and R_{3R} = 8.6kN (< 10)

With the assistance of Mathcad or a spreadsheet, the racking calculations may be rapidly assessed for both walls, including for openings.

As a note of caution, engineering judgement is needed. More experienced engineers will be able to look at a building plan and intuitively deduce the walls which offer the best stability solution.

Other simplifications might involve:

- Ignore internal stud walls *(assuming floor diaphragm is sufficient. See* **section 3.3**).
- Ignore shorter walls.
- With simplifications taken into account, allow directional wind reduction if orientation permits reduced wind load.

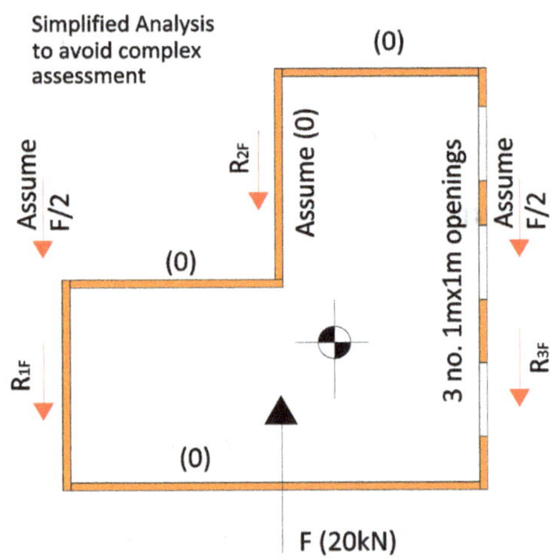

Figure 3.4.13 Simplified Racking Check

The Importance of Robust Design

A practical general approach is to work from simplified to more complex verification should the initial method not result in a satisfactory outcome. A clearly annotated 'stability plan' showing selected stabilising walls included in the calculations submitted is essential. Frequently, architects make internal wall changes or alterations to openings. This may involve client changes or planning changes at the design stage. Project budgets are constrained, particularly if the engineer is working to a fixed fee. In addition to those challenges, buildings often undergo future changes. A record of the stability walls selected enables informed assessment of proposed alterations .

3.5 WIND RESISTING FRAMES AT LARGE OPENINGS

When a building has large openings, it is most often necessary to design a frame which will resist wind loadings on the wall with openings. A typical situation where that occurs in shown in **figure 3.5.1** in which a 1 storey workshop, which is long and has a narrow width at the gable ends. The critical wind load condition acts on the long face at the eaves level, which is resisted equally at the gable ends. One gable is a solid wall. The other gable has a large opening for a door to admit vehicles. Assuming the solid wall is adequate for racking, a portal frame is built into the flanking walls as illustrated in **figure 3.5.2**, formed from a glulam lintel sitting on posts of the same width.

Often, they are called goal post frames or windframes.

For a force, 2F, at eaves level the reaction at each gable may be assumed to be F.

The lintel is checked for the ridge beam point load, assuming that it is simply supported.

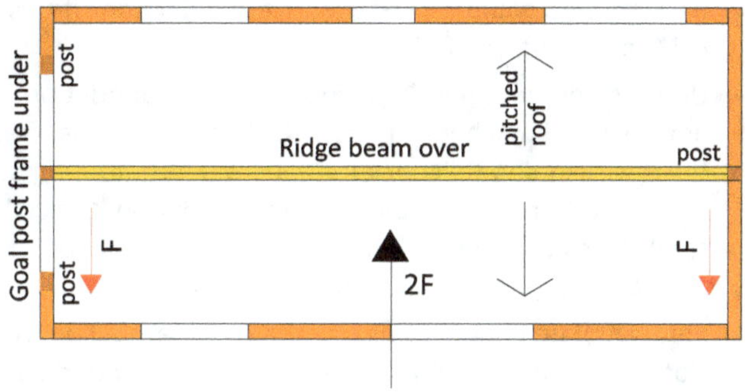

Figure 3.5.1 Workshop Plan

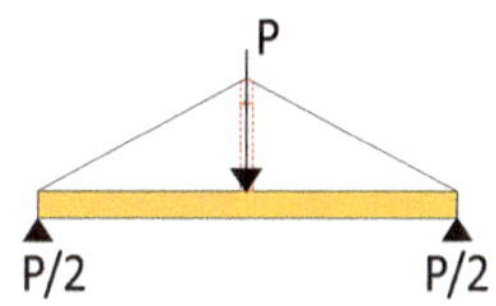

Lintel - assume simply supported. Point load from ridge.

Wind Resisting Frame (glulam/steel typical) Simplified Force /Diagram

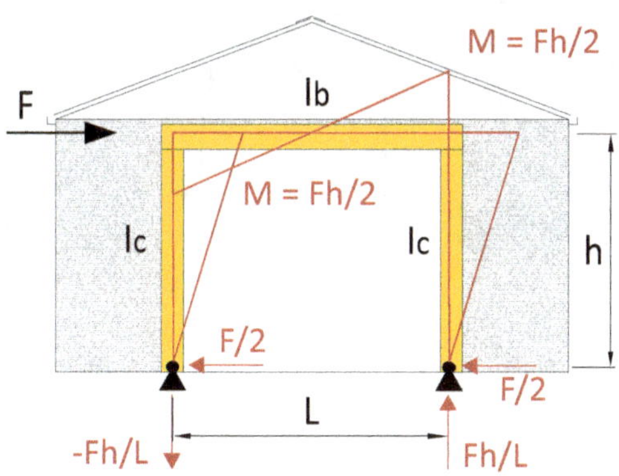

Figure 3.5.2 Workshop Entrance

It is safe to assume that the base of the frame is pinned. Fixed bases enable a more rigid frame, but must be resisted by the foundations. The simple equation below assumes a pinned base. It may be adopted for initial assessments of the frame composition and element sizes. The lintel has to take into account any vertical loads, such as from the ridge beam shown.

Forces and Lateral Deflection

Moment at Knee = Fh/2
Shear in knee connection = F/2
Shear at each base = F/2
Uplift (windward end) = -Fh/L
(check balance with dead loads)

$$\text{Eaves Deflection} = \frac{Fh^3(1/2k + 1)}{6.E.Ic}$$

Where $k = \frac{h.Ib}{L.Ic}$ *and I is inertia*

Source of equation: Steel Designers Manual and reference[5].

Comments on Wind Resisting Frames

Design considerations that need to be taken into account when evaluating portal or windframes:

- Stiffness of Frame. Normal serviceability limits for eaves lateral deflection are h/250-300 per storey (if h=2.5, Δ = 8 - 10mm). *Source: BS5950 Table 8.*
- If masonry clad, the eaves lateral serviceability limit is h/500 (if h = 2.5 Δ = 5mm).
- Stiffness assistance from flanking walls. The workshop entrance has reasonably wide walls to the side. A finite element plate (FEM) model incorporating frame and wall stiffness could indicate reduced deflections if this is of concern. Discussion of FEM modelling will follow in this chapter.
- Uplift. A concrete plinth, anchored to the foundation may be necessary. The base detail should be checked. For lighter loaded timber frames on block sleeper walls, the engineer should check for padstone requirements, anchor tension capacity and/or additional strapping to the foundation (typically thin stainless or galvanised steel straps with multiple fixings).
- Capacity of moment connections. If the loading is light and the framing is timber, large area nailplate connections may be considered. For greater moment resistance, a slotted connection with an inner steel plate with dowels or bolts can be adopted. For a steel frame, end plate moment connections are often used. See **figure 3.5.3**.
- If the frame is incorporated within the depth of a SIP or stud wall, then the frame should normally fit within the width of the SIP wall.
- Cold bridging is a consideration, particularly for steel.

Figure 3.5.3 below illustrates three types of moment connections adopted on windframes: Large nailplates when fitted to both faces of a joint (up to 6kNm capacity ultimate); internal steel plate with four or more dowels or bolts (6-20kNm capacity ultimate) and steel moment connections (10-75kNm capacity ultimate).

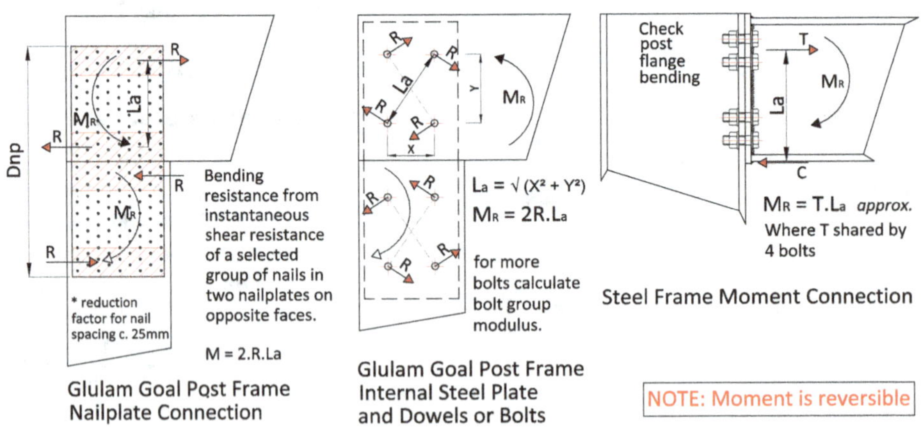

Figure 3.5.3 Typical Moment Frame Connections

3.6 FINITE ELEMENT MODELS

Finite element models with elastic isotropic properties can be very useful when evaluating walls for racking. A wall example in **figure 3.6.1** shows a SIP wall, comprising Kingspan panels with a double door on the ground floor and three windows.

Racking Check for Ground Floor Panel to BS 5268 Part 6.1

A racking check, ignoring *sliding resistance and overturning* for the purposes of the exercise, with wind and dead loading as shown in **figure 3.6.2,** gave the following racking resistance:

For a basic racking resistance of 5m length wall, R_v, of 4.3kN/m (see **section 3.2.1**), 15kN cumulative wind load applied at first floor, and overall dead load of 5kN/m:

K102 = K103 = 1.0
K104 = 2.4/2.7 = 0.89 (*height factor*)
K105 = $(L/2.4)^{0.4}$ = $(5.0/2.4)^{0.4}$ = 1.34....use 1.32 maximum (*length factor*)
K106 = $(1-1.3\rho)^2$ = 0.37 , for ρ = 0.30
K107 = K107 = $1+(0.09.w + 0.0015.w^2).(2.4/L)^{0.4}$ - *ignoring wall self weight*
= $1 + (0.09 \times 5 + 0.0015 \times 5^2) \times (2.4 / 5)^{0.40}$ = 1.36
K108 = 1.1

$R_{rk} = R_v \times L \times K_m \times K_w$ = 4.3x5x0.89x1.32x0.37x1.36x1.1 = 14.0kN < 15kN (*7% lower*)

Finite Element Racking Model

A model was compiled with two-dimensional plate elements broken down, or meshed, into elements which were approximately 0.3m x 0.3m or less. Timber studs and lintels were incorporated as beam elements, assuming C24 grade properties (nominally 110x100 section in Kingspan panels).

Base nodes were pinned, and the upper nodes restrained in the 'Y' direction into the page.

Loading was applied as drawn in **figure 3.6.2**.

The average in-plane shear forces (in kN/m) and deflections are shown in **figure 3.6.3**.

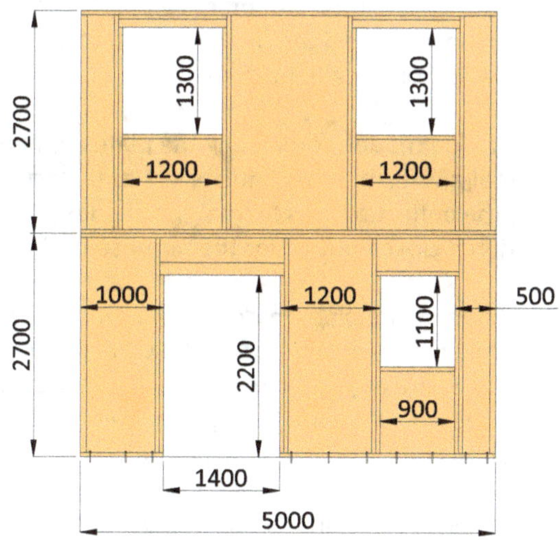

Figure 3.6.1 Racking FEM Wall Elevation

3.6.1 Summary of Finite Element Model Output

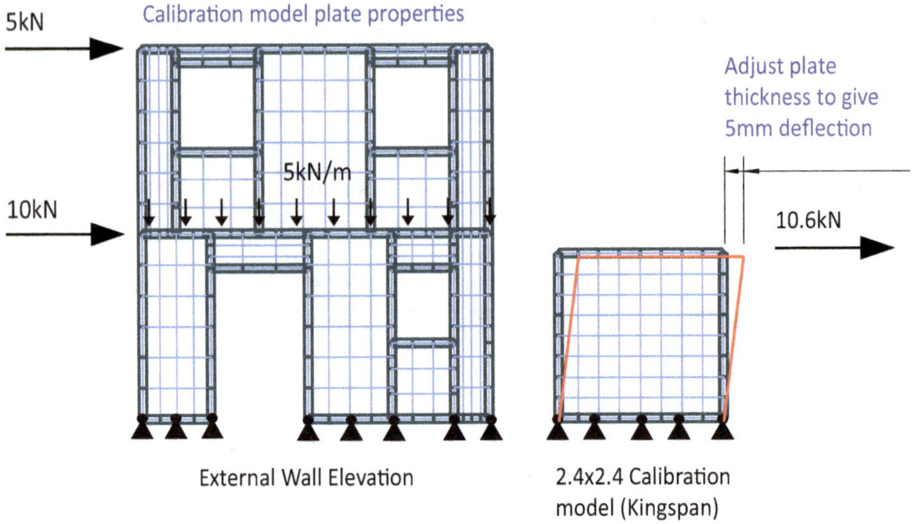

Figure 3.6.2 Plate Model Calibration and Loading

Derivation of Plate Element Properties

To evaluate plate element properties for a plate element is not a simple procedure. Given that the purpose of the model is specifically to analyse in-plane shear distribution and the estimated deflection under wind loads, a simplified procedure was to set up a 2.4mx2.4m model alongside to represent a test panel. The basic racking capacity provided by the manufacturer is based on a stiffness criterion of circa 5mm deflection for a 2.4m high panel.

A calibration model, comprising the plate elements used for the wall model, was run on a trial-and-error basis until its lateral deflection at the top was 5mm. In **figure 3.6.3,** the deflection was 5.7mm for the designated plate properties, which was considered close enough for the purpose of the model.

For reference, the plate thickness was 142mm, with effective elastic Young's modulus (E) of 30N/mm² and Poisson's ratio 0.20. It is worth noting that since a calibration model was run, any combination of properties, for instance, assuming 30mm (both faces of 15mm OSB for a Kingspan panel) thickness and corresponding modulus to yield 5mm deflection might be adopted.

With respect to the timber studs and lintels for the wall model, a long term modulus of 7250 N/mm² was assumed (11,600/1.6 - from EC5 K_{def} deformation factor). As a comparison, the model was run with short term timber modulus of 11,600 N/mm², which is shown in **figure 3.6.4**.

Comments on Model Output

The model demonstrates that the average in-plane shear of maximum 5.3kN/m is in the order of 20% greater than the published panel racking capacity of 4.3kN/m. Although this is higher, the racking check to BS 5268 part 6.1, with dead load enhancement in particular, indicates a greater local in-plane shear is not a concern.

The deflection of the wall is 5.6mm on first floor, equating to a height/deflection ratio of 480. This ratio is within 4% of the normal acceptable limit for masonry cladding of h/500.

If the ground floor wall was evaluated in accordance with PD 6693-1 method, it would have to be split into two panels in a similar way to the wall in **figure 3.2.9**. Clearly, the wall panel to the right of the door, with the window opening, would effectively have to resist the full lateral load of 15kN on first floor, since the panel to the left of the door is narrow in relation to its height. In **figure 3.6.3**, assuming the upper and lower walls being tied together as in **figure 8.3.5**, the maximum average shear over the door head is 2.9kN/m; it is reasonable to assume a 5m wall.

It is reasonable to assume sufficiency in racking for the wall from this model.

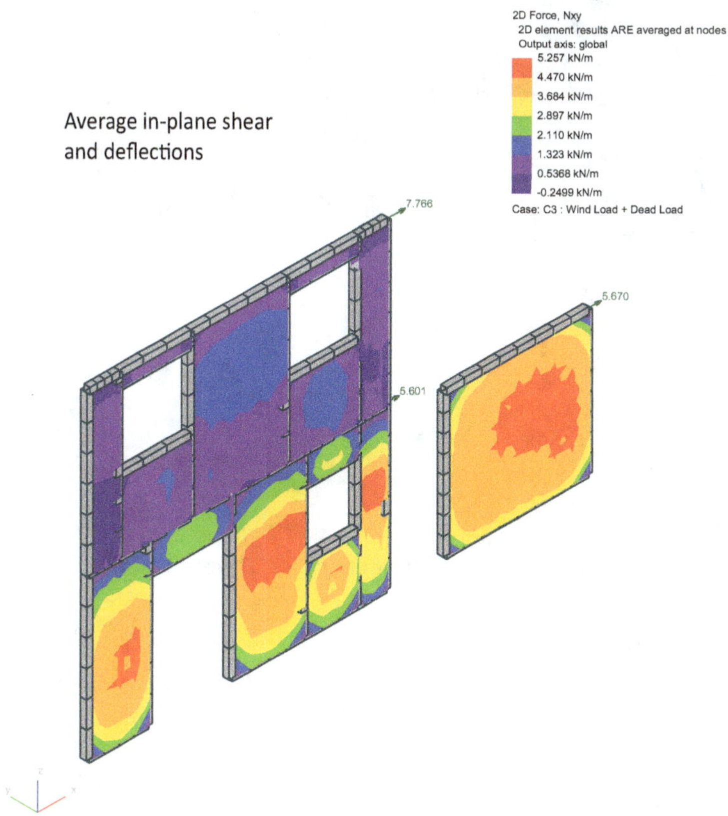

Figure 3.6.3 Model In-Plane Shear and Deflection - E Timber Long Term

Model Adjusted for Instantaneous (Short Term) Timber Modulus

In the model output in **figure 3.6.4,** an instantaneous modulus (E) of 11,600 N/mm² was used for comparison. The influence on the overall deflection and shear was small in comparison with **figure 3.6.3**. The overall reduction in deflection to 5mm on first floor equates to a ratio of h/540, and the in plane shear average, maximum 4.7kN/m is reduced from the 5.3kN/m in the former model. It should also be borne in mind that tested SIP panels fail at shear levels in excess of 15-20kN/m - although for a given manufacturer, the designer should satisfy themselves that test data confirm this.

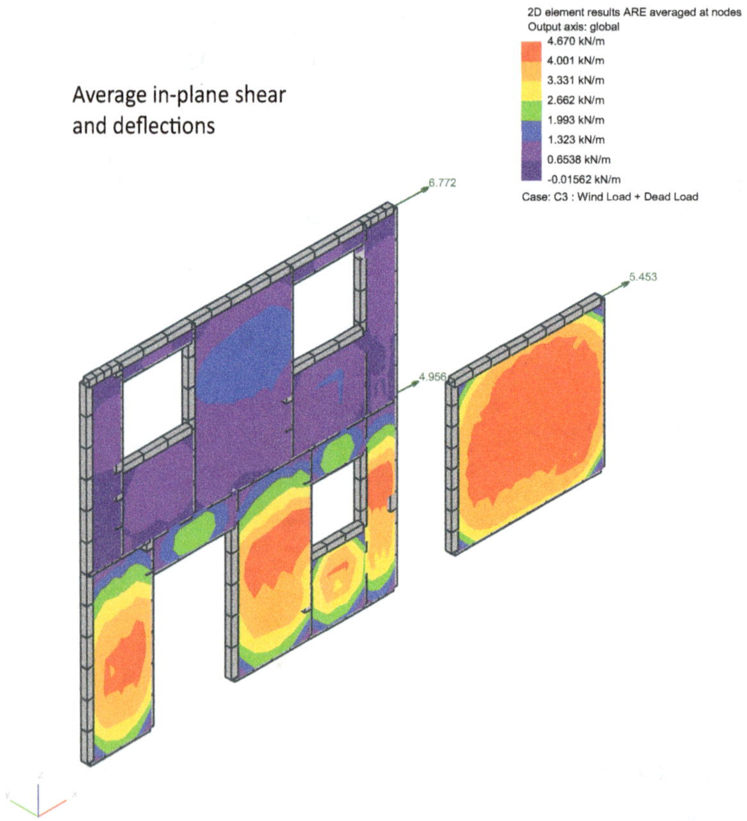

Figure 3.6.4 Model In-Plane Shear and Deflection - E Timber Instantaneous

Notes on Overturning and Sliding

There was no overturning when checked by the method outlined in **section 3.2.6**. However, the restoring and overturning moments were close to parity. If the wind load was higher, or the dead load on the wall was lower, provision may have to be made for strapping down or anchorage at both ends of the wall. The model shows a shear at the base of circa 4kN/m with a relatively low dead load, hence low friction resistance. In this case, the sole plate anchors need careful checks, which might necessitate additional straps. Straps are shown in chapter 8, **figure 8.3.5**, with reference to connections and fixings.

3.6.2 Other Applications of FEM Models

Apart from the advantage of being able to examine specific walls as a design tool for assessing the influence of racking loads, FEM models can be useful for other purposes.

Interaction of Bracing or Windframes with SIP walls

The stiffening effect of SIP panels may be evaluated, as in the restrained gable shown in **figure 3.5.2** .The flanking walls to the side of the frame will stiffen the windframe and share some of the lateral load.

Monocoque Garden Office and Modular Units

Whole models of modular units can be very useful when verifying stability, foundation load distribution and lifting deformations of prefabricated units when hoisted into site locations. See **figure 7.3.1** showing an example of a garden room.

Progressive Collapse or Disproportionate Collapse Assessments

Schools, hotels, residential blocks and public buildings are subject to disproportionate collapse checks. Methods include the removal of a key structural element or a substantial part of a lower wall to demonstrate structural integrity. In **figure 3.6.5,** the ground floor panel was removed. The first floor wall supports dead and live loads from first floor and roof with appropriate 'accidental' factors. Deflection is not significant. The SIP panels peak in-plane shear was 6kN/m, and the corner post compression of 24kN with an effective length of 2.7m were both found to be acceptable.

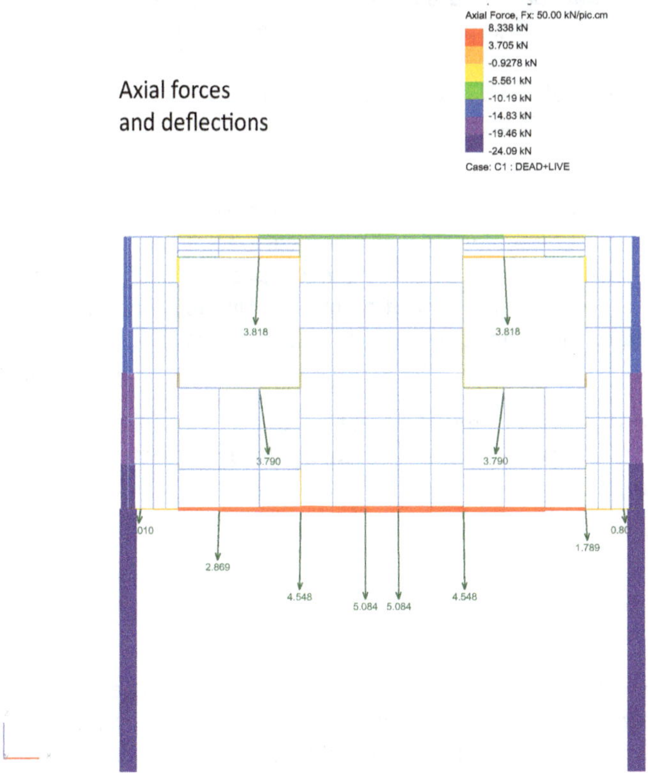

Figure 3.6.5 Progressive Collapse Model

Chapter 3

Quick Note Summary

Racking (In-Plane Shear)

Wind loads or earthquakes cause horizontally applied (or lateral) forces to act on the external walls of buildings. The capacity of the framing must be checked to ensure that the building is capable of resisting lateral loading by means of racking resistance in walls.

Overall stability may be achieved solely by external wall SIP panels or in combination with internal sheathed stud or SIP walls. Timber frame construction in the UK is commonly platform frame construction, with joists separating upper and lower walls. SIP construction, of external walls in particular, is commonly constructed with interconnection of walls separated by the floor deck, with joists fixed to the side of the lower wall.

Calculation Methods

In the UK, design document PD-6693-1 is generally advocated for platform frame construction of timber frame walls. BS 5268 Part 6.1 employs permissible stress analysis, not specific to wall anchorage. Eurocode 5 comprises methods A and B. Method A is specific to walls anchored at the ends. Method B is derived from BS 5268 part 6.1 but has inaccuracies in the interpretation of wall dimensional factors.

BS 5268 Part 6.1 is the method followed in the book, consistent with SIP manufacturers tested SIP panel racking resistances.

Racking Checks

Verification of panels for racking resistance comprises three parts: sliding, overturning and racking strength. Presence of openings in walls, vertical loading and dimensional configuration all affect the panel capacity. Variations in walls compared from standard test panels are taken into account with modification factors. Distribution of racking loads in buildings is related to the relative stiffness of stabilising walls.

Floor Diaphragms

Decking for floors must be capable of transmitting lateral loads to stabilising walls. This is generally satisfactory for standard construction, but local checks may be necessary around floor voids.

Supplementary Structural Framing

Where large openings occur, it may be necessary to provide additional stability with structural framing, commonly constructed in sawn timber, glulam or steel.

Finite Element Analysis

2D finite element plate models are useful for modelling SIP wall assemblies, particularly when assessing the combined racking of two or more storey walls. The stiffening influence may be represented for SIP walls alone, or combined with portal frames and for progressive collapse where lower walls are removed to represent failure or collapse of supporting walls under.

Chapter **4**

Axial Compression (Walls)

Compressive loads acting on OSB SIPs panels. Critical load assessment: compressive and buckling. Eccentricity. Test data. Influence of timber splines. Combined compression and bending. Stud walls. Panels sitting on walls which are more narrow. Panels with precast floors. Examples of MgO panels.

4.1 AXIAL COMPRESSION LOADING

SIPs panel suppliers and installers generally have warranties which cover up to three storey buildings. The vertical loading on walls for buildings up to this height, being restrained at each floor, are rarely greater than circa 30-40kN/m for unfactored combined dead and live loading. Calculation methods and test data presented in this chapter demonstrate that checks are not critical for most common types of building.

When walls are two unrestrained stories in height, approximating to 5.5m, panels need to be checked for load bearing resistance, which may become a problem for wind loads and roof or floor loads acting together. Compression and bending needs to be verified in such circumstances.

Hotels have been constructed with precast floors sitting on SIP party walls which present greater dead and live loading when compared with domestic or school buildings with timber floors. In these circumstances, it may be necessary to reinforce the walls with internal timber or steel posts. See **figure 4.9.1**.

School assembly rooms and sports halls, for example, have long span roofs with framing, which will need to be incorporated with support posts, which may have to provide portal action for stability. SIP walls built between posts will only have limited vertical loading to support, their main structural function being to provide lateral stability to the frames.

Guidance for the design of SIP walls is limited. The method for analysis in compression presented in this chapter was taken from Allen[2], in combination with the secant formula for eccentric loading if applicable. It will be shown to give results which have a safe margin when compared with available test data. Eurocode 5 gives a method for checking compression and bending combined. Internal stud wall checks are also addressed.

Figure 4.1.1 Panel in Compression

Stud walls, in **section 4.7**, and panels with MgO facing, in **section 4.10**, will also be reviewed with typical properties and a calculation example to follow.

4.1.1 Panel Behaviour Under Axial Compression Loads

For short panels, circa 1.5m and less in height, compression resistance is generally governed by compressive resistance of the faces. For walls that are greater in height, therefore, more slender, buckling becomes more critical. Verification involves checking for both compression resistance in the faces and critical applied load from which the lower value is selected for the design.

Derivations of Euler or Rankine formulae may be employed for buckling analyses, which take into account the combined bending stiffness of faces and the shear stiffness of the core.

In the classic text book 'Analysis and Design of Structural Sandwich Panels' by Howard Allen[2], a formula for buckling will be presented, which takes into account the interaction of core shear and face bending. It was found to give satisfactory results without undue complication. The secant formula, from Roark[14], for deriving buckling under eccentric loading may also be included with this approach.

Compression Modes of Failure

From the paper by Rungthonkit[6] three modes of failure under a compressive load are described: bulging of the skin, peeling failure of the skin, pulling away from the core at header or soleplate and buckling.

2.44m high, 125mm SBS panels, were tested to failure[6] under a compressive load. Both failed by end bearing, where the OSB faces at the base of at least one of the panels delaminated. The OSB faces indicated a 'ragged' edge after testing.

In other tests on SBS and Hemsec panels, which are very similar, with PUR cores and OSB/3 faces, failure was by a variety of modes: end bearing, delamination (peeling) and buckling of one or both faces, within the panel. See **figure 4.1.2**.

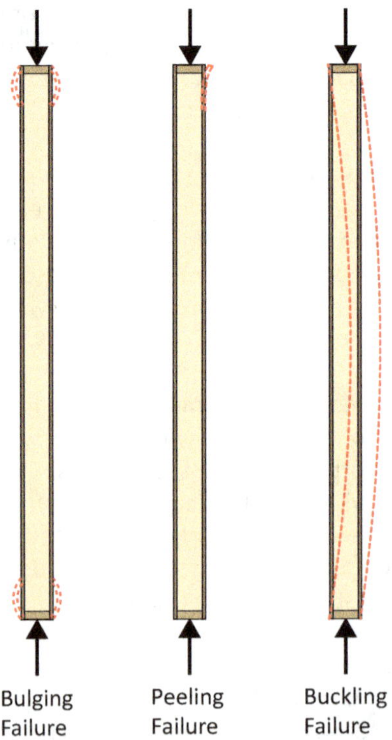

Figure 4.1.2 Compression Failure Modes

Results from the tests are shown in **table 4.5.1**. It shows average values at failure compared with values predicted by calculation, based on methods of evaluation presented in this chapter.

It can be seen from the table that the Allen[2] method of calculation for evaluating design resistances, including for the effects of eccentric loading and Eurocode 5 long term load and material factors, are all lower than tested values by a comfortable margin.

4.2 AXIAL COMPRESSION BEARING FORMULA

As noted in **section 4.1.1** verification of panels entails the selection of the lower value for compression resistance based on two separate checks: bearing resistance on OSB faces and critical buckling load. In this section, bearing resistance is considered.

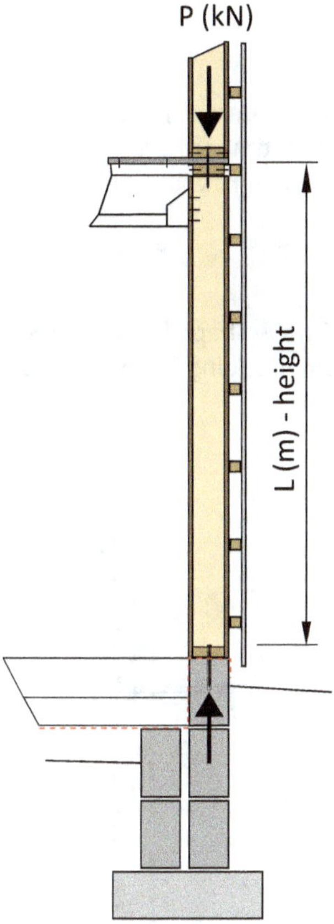

Figure 4.2.1 Axially Loaded Wall

Bearing Resistance, P_b

In **figure 4.2.2,** a short panel, in which buckling failure is not critical owing to its low slenderness, will have a resistance in compression, which is governed by the compressive strength of the OSB faces parallel to the grain. The applied load, P, is resisted equally by each face.

Characteristic bearing resistance, P_b, is given by:

$P_b = 2.f_{c,0,k}.B_p.t$ ...where

$f_{c,0,k}$ = Characteristic compression strength of OSB parallel to the grain (stress in N/mm²)

B_p = panel width
t = face thickness
assuming both faces have equal thickness.

Note that in the example, a single soleplate sits on the supporting block, which is the width of the panel, with full bearing on the faces. If the SIP panel is wider than the support, then a wider base soleplate as shown in **figure 8.3.3** would act as a load spreader ensuring each face is supported.

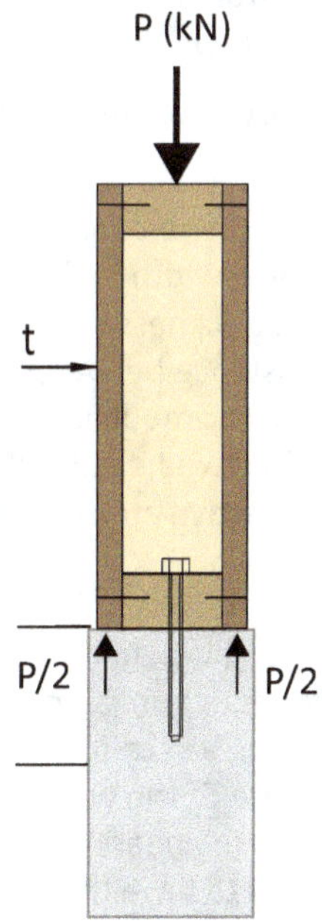

Figure 4.2.2 Face Bearing

Characteristic Bearing Resistance Example

A 1m high panel with 11mm faces, with $f_{c,0,k}$ = 15 N/mm², sits fully on its support:

$P_b = 2 \times 15 \times 1000 \times 11 = 330\times10^3$ N (330kN/m)

The design resistance to Eurocode 5 is found overleaf.

Design Bearing Resistance Example (from the Previous Page)

From Eurocode 5, assuming a load duration factor for long term, service class 2, $k_{mod(OSB)} = 0.40$, from **section 2.2.1**. The partial material factor, $\gamma_M = 1.2$.
Design bearing resistance, $P_{b,d}$ is calculated below:

$$P_{b,d} = \frac{P_b \cdot k_{mod(osb)}}{\gamma_M} = 330 \times 0.40/1.2 = 110kN/m$$

This is seldom exceeded in most 2-3 storey domestic and commercial properties. For example, a 3 storey school building with classrooms with timber floor widths' contribution 2.5m, and roof contribution 2.5m, with 8.25m height walls, timber clad would have an ultimate line load on the wall of circa 40kN/m.

4.3 ALLEN[2] COMPRESSION BUCKLING FORMULA

From Allen[2] the buckling critical load P_{cr}, which has a half wave shape in **figure 4.3.1**, assumed for a 'strut' with pinned supports top and bottom, buckling factor, k = 1.0, is calculated from the formula:

$$P_{cr} = \frac{P_E}{1 + \frac{P_E}{A_{eff}.G_{c(B)}}}$$ ..where Euler buckling load, $P_E = \frac{\pi^2.D}{L_{eff}^2}$

and flexural rigidity $D = E_{f(0.05)}.\frac{B_p.t.d_{eff}^2}{2}$

The rigidity is based on the inertia of the faces about the neutral axis. Bending stiffness of the faces and core are negligible in comparison.

Summary of terms:

$E_{f(0.05)} = (0.85.E_{(mean)})$ *from Medite Smartply datasheet*
...*fifth percentile* face modulus of elasticity
B_p = panel width
t = face thickness
d_{eff} = distance between centre of faces ($d_{eff} = d_p - t$)
$G_{c(B)}$ = shear modulus of the core from **section 2.3**
L_{eff} = Effective height (L_{eff} = L, for k = 1.0)
A_{eff} = Effective area of foam core = $B_p.d_{eff}$

Of note, this assumes no eccentricity of load from the floor joists. In the next section, a derivation for a load reduction factor, based on the eccentricity of load using a secant formula, will review the reduced resistance of the wall with eccentric loads. This also assumes full bearing on the supporting wall which is assumed to have the same or greater width than the wall panel. If the wall is narrower or wider than the wall above, then eccentricity of loading should be considered.

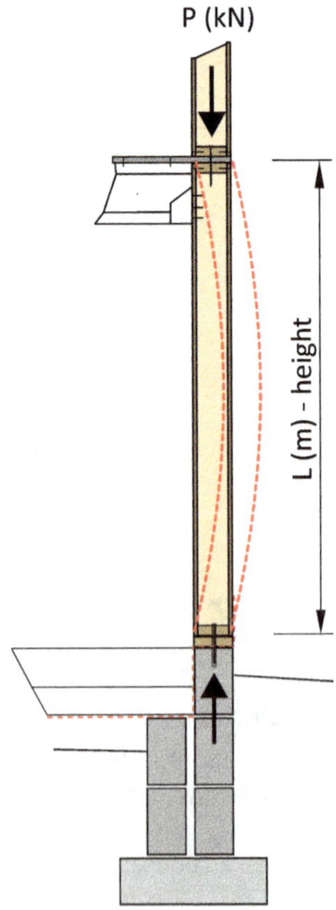

Figure 4.3.1 Axial Buckling

Critical Buckling Check Example

From the previous page, in the example of **figure 4.3.2,** a 125mm wide wall panel, 3m high, has the following parameters:

$E_{f(0.05)}$ = modulus of elasticity of the faces = 3800 x 0.85 = 3230 N/mm^2
Note: $E_{f(mean)}$ = 3800, then $E_{(0.05)}$ is obtained using a factor of 0.85
B_p = panel width = 1.2m
t = face thickness = 11mm
d_{eff} = distance between centre of faces ($d_{eff} = d_p - t$)
= 125 - 11 = 114mm
G = shear modulus of the core = 2.3 N/mm^2
L_{eff} = Effective height (L_{eff} = L, for k = 1.0) = 3m
A_{eff} = Effective area of foam core = $B_p.d_{eff}$
= 1200 x 114 = 136,800mm^2

Calculation of the critical load, P_{cr} is as follows:

Flexural rigidity, $D = E_f.I_f$
= 3230 x 1200 x 11 x 114^2/2 = 277.0x10^9 Nmm^2

Euler buckling, critical load, $P_E = \pi^2.D/L_{eff}^2$
= 3.142^2 x 277 x10^9 / 3000^2 /1000 = 304kN

Therefore, P_{cr}, can be calculated:

$$P_{cr} = \frac{P_E}{1 + \dfrac{P_E}{A_{eff}.G_{c(B)}}}$$

Applying consistent units:
P_{cr} = 304 x10^3/ (1 + 304 x 10^3/136,800/2.3)/1000
= 155kN

This result may be compared with the tested average result obtained from 'CERAM453a.05' in **table 4.5.1,** which follows in the chapter, at 285kN. The calculated P_{cr} value is circa 60% of the test value.

Other values give a similar result. For instance, the 5.5m high wall, 'CERAM091127' has a test value of 133kN and calculated P_{cr} of 80kN (circa 60% of the test result).

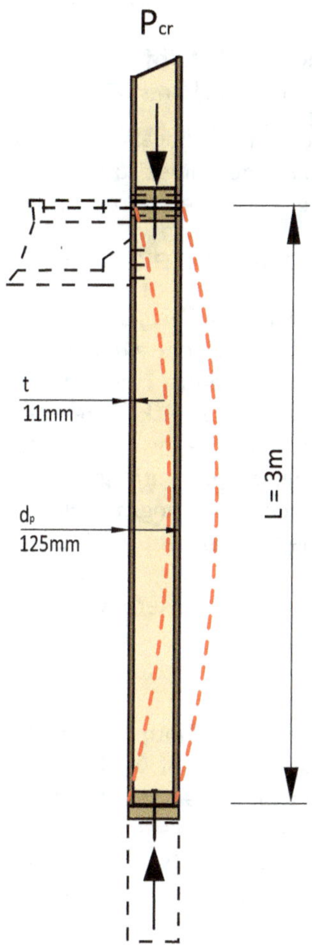

Figure 4.3.2 Axial Buckling Example

From the Rungthonkit paper[6] the test results are almost parity with the calculated value. It is not known why this is the case since the CERAM and Lucideon testing laboratories, which were independent tests, show consistently higher failure values. Nevertheless, the calculated values are all within test margins. Design buckling resistances assure a further safety margin.

Critical Buckling Design Resistances

From Eurocode 5, assuming a load duration factor for a *long term*, service class 2, $k_{mod(OSB)}$ = 0.40 and partial material factor, γ_M of 1.2, then design buckling resistance, for a 1m width, $P_{cr,d}$ is calculated below:

$$P_{cr,d} = \frac{P_{cr} \cdot k_{mod(osb)}}{\gamma_M} = 155 \times 0.40/1.2/1.2 = 43\text{kN/m}$$

Comparison of Bearing and Buckling Design Resistances

From **section 4.2** the bearing design resistance is compared with the buckling value:

$P_{b,d}$ = 110kN/m
$P_{cr,d}$ = 43kN/m < 110 (*critical*)

Buckling is critical for the 3m high wall. The lower value must, therefore, be applied as the design resistance. This would still be greater than the 40kN/m ultimate loading from the three storey school described in **section 4.2**.

4.4 ECCENTRIC BUCKLING FORMULA

When a load acts eccentrically on a wall - in other words, the centre of loading does not correspond to the wall centre - this induces a moment which further reduces the buckling resistance. **Figure 4.4.1** shows a wall where the resultant load from the joists and wall acts off centre. A secant formula from Roark[14] for eccentric loading is applied to give a coefficient, K_e, for the calculated buckling load assuming no eccentricity:

$$K_e = 1 + \frac{e.\bar{y}}{r_y^2} \cdot \sec\left\{\frac{k.L}{2r_y} \cdot \sqrt{\left(\frac{P}{A_f.E_{f(0.05)}}\right)}\right\}$$

Where....
A_f = Area of both faces = $2.B_p.t$
B_p = wall/panel width
E_f = Modulus of elasticity of face
$\bar{y}$ = extreme distance of face from neutral axis = $d_p/2$
e = eccentricity from centre of wall
r_y = radius of gyration based on faces = $\sqrt{(I_f/A_f)}$
I_f = Moment of inertia of faces = $(d_p^3 - d_c^3).B_p/12$
k = factor based on end restraints
(k = 1.0 in **figure 4.4.1**)

Modified critical buckling load P_{cre}, is the product of the secant formula result and the calculated critical buckling load assuming no eccentricity:

$P_{cre} = P_{cr} / K_e$

An example follows, based on the school building described in **section 4.3**. with 172mm walls.

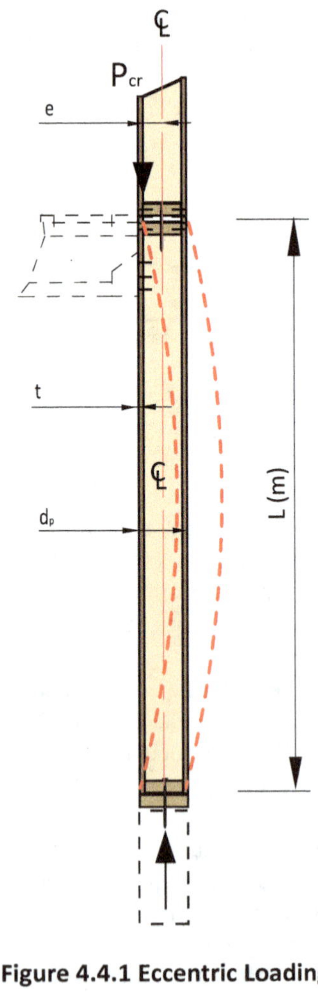

Figure 4.4.1 Eccentric Loading

Critical Buckling Check Example

In the example of **figure 4.4.2,** a 172mm wall panel, 3m high, has loading assumed to act in the centre of the wall. It has the following parameters for a 1m strip of wall:

Where....

A_f = Area of both faces = $2.B_p.t = 2 \times 1000 \times 15 = 30{,}000\text{mm}^2$

B_p = wall/panel width = 1m (1000mm)

$E_{f(0.05)}$ = fifth percentile modulus of elasticity of face = $0.85 \times 3800 = 3230\text{N/mm}^2$

$\bar{y}$ = extreme distance of face from neutral axis = $d_p/2 = 172/2 = 86\text{mm}$

e = eccentricity from centre of wall - see calculation in **figure 4.4.3** and below.

I_f = Moment of inertia of faces $= (d_p^3 - d_c^3).B_p/12 = \{172^3-(172-2\times15)^3\}\times1000/12$
$= 185.4\times10^6\ \text{mm}^4$

r_y = radius of gyration based on faces
$= \sqrt{(I_f/A_f)} = \sqrt{(185.4\times10^6/30\times10^3)} = 78.6\text{mm}$

For simplicity dead and live loading are combined.
If roof loading, F_r = 4kN/m, wall loading (2 stories), F_w = 6kN/m and joist loading, F_j = 6kN/m then, from **figure 4.4.3**:

$e = ((4+6) \times 86 + 6 \times 172) / (4+6+6) - 86 = 32\text{mm}$

Total load for eccentric moment, $P = 1.5\times(4+6+6)$
= 24kN/m ...*1.5 ultimate factor assumed conservatively*

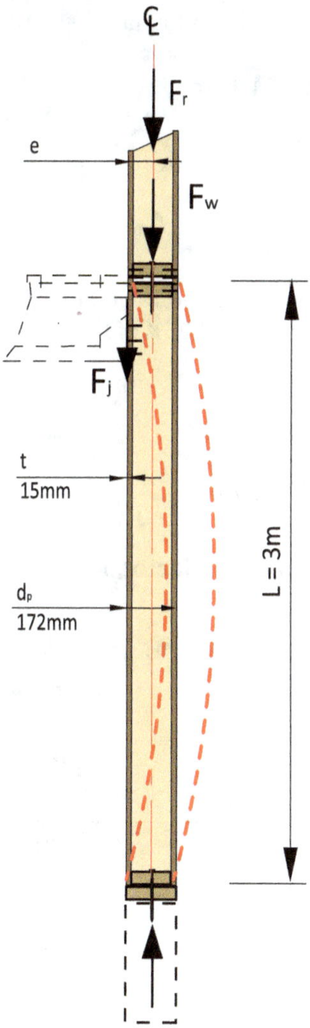

Figure 4.4.2 Eccentric Loading Example

Calculate from right face:

$$e = \frac{(F_r + F_w).dp/2 + F_j.d_p - d_p/2}{F_r + F_w + F_j}$$

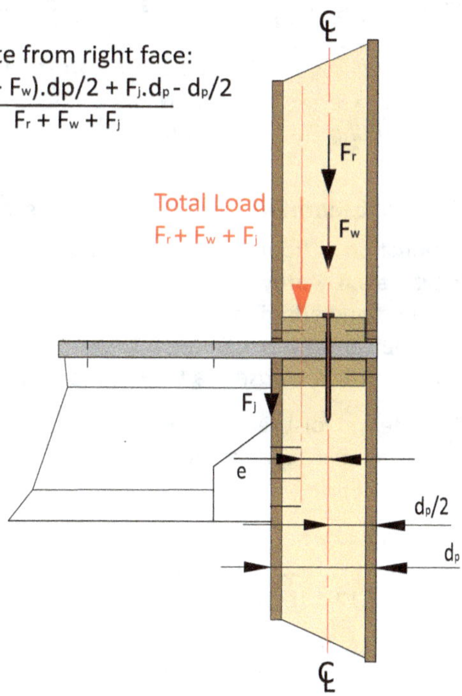

Figure 4.4.3 Eccentricity Location Calculation

Calculate Eccentricity Coefficient, K_e

Recalling the formula for the coefficient, or reduction factor for eccentricity, K_e:

$$K_e = 1 + \frac{e.\bar{y}}{r_y^2} \cdot \sec\left\{\frac{k.L}{2r_y} \cdot \sqrt{\left(\frac{P}{A_f.E_f}\right)}\right\}$$ on the assumption k = 1.0 ($P = 24\text{x}10^3$N)

$K_e = 1 + 32\text{x}86/78.6^2 \text{ x } \sec \{(1\text{x}3000/2/78.6)\text{x} \sqrt{(24\text{x}10^3/30\text{x}10^3/3230)}\}$
$= 1 + 0.445 \text{ x } \sec (19.1 \text{ x } 0.0157) = 1 + 0.445\text{x}1.05$ - *note* $\sec\theta = 1/\cos\theta$, *θ in radians*
$= 1.47$

Note: $1/K_e = 0.68$

If the critical buckling load, using the method in **section 4.3**, P_{cr} = 234kN/m, then the critical *design* long term load, including for eccentricity, P_{cre} can be found:

$$P_{cre,d} = \frac{P_{cre} \cdot k_{mod(osb)}}{\gamma_M} \cdot \frac{1}{k_e} = \frac{234 \text{ x } 0.40}{1.2} \text{ x } \frac{1}{1.47} = 53\text{kN/m} > 24\text{kN/m}$$

The panel is adequate for the loading applied.

Simplification of Calculation for P_{cre}

To simplify the calculation, for estimation purposes, if the calculated value of P_{cr} is substituted in the equation for K_e, and an assumed eccentricity of 25% of the panel width is assumed (43mm), the values obtained are:

$K_e = 1.63$ ($1/K_e = 0.62$), therefore...pro-rata...
$P_{cre,d} = 53 \text{ x } 1.47/1.63 = 48\text{kN/m} > 24\text{kN/m}$
The resulting P_{cre} is 91% of the value calculated above (53kN/m).

Eccentric Moment Resistance by Joists and Floor (Reduced factor, k)

Connection of floor joists for the structure shown in **figure 4.4.4** provides resistance to the eccentric moment and additional wall restraint. It can be argued that the connection can resist the eccentric moment, *which would increase the buckling resistance.*

For a design joist reaction, F = 2.5kN , and eccentricity of 45mm, this exerts a couple in the joist of effective depth 150mm.

Force in flanges = 2.5 x 0.045/0.15 = ±0.75kN
If the hanger face is 100x100mm, then referring to **section 2.12.1** resistance of the foam core, if EPS100:

$P_{pd,EPS} \approx$ 1kN > 0.75

Top tension can be resisted by shear in the deck fixings. The wall resistance could assume that there is no eccentricity. Also, for the wall, k ≈ 0.85 ($L_{eff} < L$).

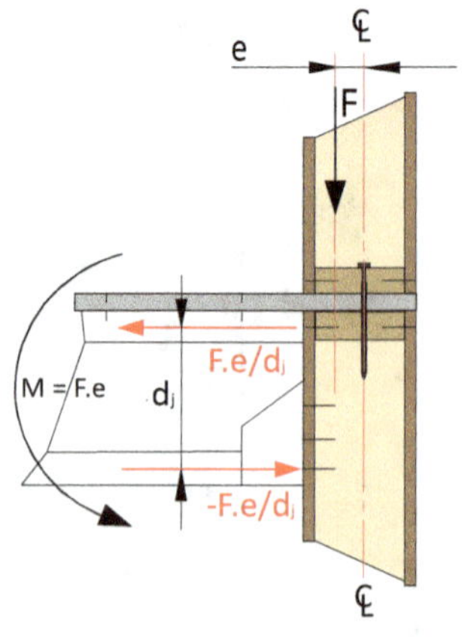

Figure 4.4.4 Eccentric Load Joist Restraint Mechanism

4.5 COMPARISON OF TEST DATA WITH CALCULATED AXIAL RESISTANCES

From available test data for compressive axial loads with and without eccentricities, calculations were undertaken in accordance with the methods outlined in **sections 4.2, 4.3 and 4.4**. Calculated characteristic values, which *exclude load duration and material factors,* are shown to the right of the test failures below in **table 4.5.1**.

With the exception of the Rungthonkit[6] tests conducted at Birmingham University, the calculated values are consistently lower than the test values, ranging from circa 50-65% discrepancy. It is not known why the Rungthonkit results were close to the calculated values or why the addition of a timber spline did not improve the resistance of the 125mm panels. It was quoted that: "*A possible explanation for this is that due to the dimensional tolerance of facial panels, bearing ends do not provide a flush bearing surface, and therefore it may be likely that only part of facial panels have endured the bearing load. This causes the panel specimen to fail due to the end bearing.*"

On the other hand, the test failure values by CERAM and Lucideon were consistently higher than calculated values; included in the table are average failures from a number of test panels.

PANEL TEST SOURCE	L (m) height	Bp (m) (width)	Dp (mm) (depth)	Test Failure (kN)	Calc'd Failure (kN)	Ecc'y (mm)	Test Failure (kN)	Calc'd Failure (kN)
Rungthonkit[6] (SMC)	2.44	1.2	125mm	209kN$_1$* *174kN/m*	189kN *158kN/m*	N/A		
Rungthonkit[6] (SDC)	2.44	1.2	125mm	205kN$_1$* *170kN/m*	(---)	N/A		
CERAM352.04 SBS Ltd.	2.4	2.4	100mm	459kN$_3$** *191kN/m*	262kN *109kN/m*	25mm	310kN$_3$** *129kN/m*	160kN *67kN/m*
CERAM352.04 SBS Ltd.	2.7	2.4	100mm	366kN$_3$** *152kN/m*	237kN *99kN/m*	25mm	261kN$_3$** *109kN/m*	145kN *60kN/m*
CERAM352.04 SBS Ltd.	3.0	2.4	100mm	392kN$_3$** *163kN/m*	211kN *88kN/m*	25mm	316kN$_3$** *132kN/m*	129kN *54kN/m*
CERAM453a.05 Hemsec	3.0	1.2	125mm	285kN$_3$˫ *238kN/m*	155kN *129kN/m*	21.5mm	167kN$_3$˫ *139kN/m*	109kN *91kN/m*
CERAM091127 SBS Ltd.	5.5	1.2	125mm	133kN$_3$˫ *137kN/m*	70kN *58kN/m*	N/A		
CERAM153.03 SIPtec. Ltd	2.4	1.2	100mm	225kN$_6$* *190kN/m*	133kN *111kN/m*	25mm	123kN$_5$* *102kN/m*	81kN *68kN/m*
Lucideon171066 SBS Ltd.	3.6	4.89	75mm	454kN$_1$* *92kN/m*	197kN *40kN/m*	N/A		

Table 4.5.1 Comparison of Axial Tests in Compression with Calculated Values

Notes: **End bearing (top or bottom)/splitting of OSB from header/soleplate.*
***Delamination of OSB from core. ˫Buckling of panel.*
'**392kN$_3$***' *- denotes average failure value from 3 test samples. End bearing failure.*

Comments on Table 4.5.1 Comparison of Tested with Calculated Resistances

The bearing, buckling and eccentricity formulae presented in this chapter are consistently within the average test values in **table 4.5.1**. Confidence can be gained that the formula is applicable in practice. Further comments of note are:

- Average values were presented from the reports. Where more than one test was conducted, the following ***minimum*** results were: CERAM352.04 (2.4 high) 448kN and 297kN (eccentric); CERAM352.04 (2.7 high) 275kN and 248kN (eccentric); CERAM352.04 (3.0 high) 387kN and 308kN (eccentric); CERAM453a.05 (3.0 high) 275kN and 165kN (eccentric); CERAM153.03 (2.4 high) 214kN and 115kN (eccentric). All minimum results were greater than the calculated values.
- Owing to the small number of tests, it is not straightforward to generate a statistical test figure for comparison. From CERAM153.03, assuming a 3.3 sigma variation from the mean (98% confidence level), purely axial tests gave a standard deviation (Sd) of 8.7. Subtracting 3.3xSd from the average yields 196kN (> 133kN calculated), and for the five eccentric tests Sd = 5.3 (3.7 sigma), giving 104kN (> 81kN) for the same calculation. In short the calculations are within this criterion.
- Failure, in the case of buckling, was sudden, owing to brittle failure of one or more of the faces. The calculated values are taken as ***characteristic***, rather than design, as a more conservative approach.
- All tests were carried out without the addition of joists or base restraints. The effective lengths of panels were, therefore, assumed to be equivalent to the test panel height (k = 1.0).

4.6 ASSESSMENT OF PANELS WITH TIMBER SPLINES

The addition of timber splines to panels increases the buckling resistance. As noted in **section 4.5**, with reference to Rungthonkit[6] finite element models predicted 237kN critical buckling load, P_{cr}, for a 255mm panel with SIP spline and 307kN with a 47x103 deep C16 grade spline; the timber spline adding 70kN in resistance (20% increase).

Based on the Allen[2] formula in **section 4.3**, a formula for the addition of a timber spline is presented below. See **figure 4.6.1**.

$$P_{cr(s)} = \frac{P_E}{1 + \dfrac{P_E}{A_{eff}.G_{c(B)}}} + \frac{1}{M_r} \cdot \frac{\pi^2 E_{s(0.05)}.I_s}{L_{eff}^2} \cdot \frac{B_p}{S_s} \quad \text{...where....}$$

$P_E = \pi^2 D/L_{eff}^2$ and $D = \dfrac{E_{f(0.05)}.B_p.t.d_{eff}^2}{2}$ for the panel as in **section 4.3**.

$E_{s(0.05)}$ = Modulus parallel to the grain for *one* stud, depending on the grade of timber.
I_s = moment of inertia of *one* stud ($I_s = B_s.d_s^3 / 12$).
S_s = Stud spacing and B_p = panel width
M_r = Modular ratio of face to stud ($M_r = E_{s(0.05)}/E_{f(0.05)}$), depending on the grade of stud.

Eccentric Loading

For eccentric loading, the secant formula from **section 4.4** may be applied. The stud properties may be included as a composite section, but in most instances, the face critical buckling load predominates. The timber properties may be ignored.

Critical Buckling Check Example with Timber Splines (at 1.2m Centres)
A 142mm wall with 15mm face thickness comprises 1.2m wide panels with C24 timber splines in each joint 100x110mm deep. It is 3m in height. From **section 4.3**:

Panel Critical Buckling Resistance, P_{cr}

$E_{f(0.05)}$ = fifth percentile modulus of elasticity of the faces = 3800 x 0.85 = 3230N/mm²
B_p = panel width = 1.2m, S_s = 1.2m (B_p/S_s = 1.0)
t = face thickness = 15mm
d_{eff} = distance between centre of faces ($d_{eff} = d_p - t$) = 142 - 15 = 127mm
$G_{c(B)}$ = flexural shear modulus of the core = 2.3 N/mm²
L = Effective height (L, for k = 1.0) = 3m
A_{eff} = Effective area of foam core = $B_p.d_{eff}$
= 1200 x 127 = 152,400mm²

Flexural rigidity, D = $E_{f(0.05)}.I_f$
= 3230 x 1200 x 15 x 127²/2 = 469x10⁹ Nmm²

Euler buckling, critical load, $P_E = \pi^2.D/L^2$
= 3.142² x 469x10⁹ / 3000² /1000 = 514kN

Applying consistent units:
P_{cr} = 514 x10³ / (1 + 514 x 10³/152,400/2.3) / 1000 = 208kN

Stud Critical Buckling Resistance, P_{crs}

$$P_{crs} = \frac{1}{M_r} \cdot \frac{\pi^2 E_{s(0.05)}.I_s}{L_{eff}^2} \cdot \frac{B_p}{S_s} \quad \text{...where....}$$

$E_s = E_{0(0.05)}$ (C24) = 7400 N/mm² *from EN 338:2003*
$I_s = B_s.d_s^3/12$ = 100 x 110³ /12 = 11.1 x 10⁶ N/mm²
$M_r = E_{s(0.05)}/E_{f(0.05)}$ = 7400/3230 = 2.29
B_p/S_s = 1.2m/1.2m = 1.0

P_{crs} = (1/2.29) x 3.142² x 11.1x10⁶ x 7400 x (1.0) / 3000² /1000 = 39.6kN ≈ 40kN

Total buckling resistance, $P_{cr(s)}$ = 208 + 40 = 248kN

This was checked against the Eurocode method for studs outlined in ***section 4.8****, which gave a value of 31N (<40) for a stud. As the design k_{mod} factor is adopted for OSB/3 below (rather than C24 timber for the stud), the composite buckling resistance is considered to be conservative.*

Design buckling, resistance (long term), converted for ***a 1m strip***:

$P_{cr,s,d.1m} = P_{cr(s)} \cdot k_{mod(osb)} / \gamma_M / B_p$ = 248 x 0.40 / 1.2 / 1.2 = 69kN/m

Adding the studs increases the buckling resistance by circa 20% for this panel.

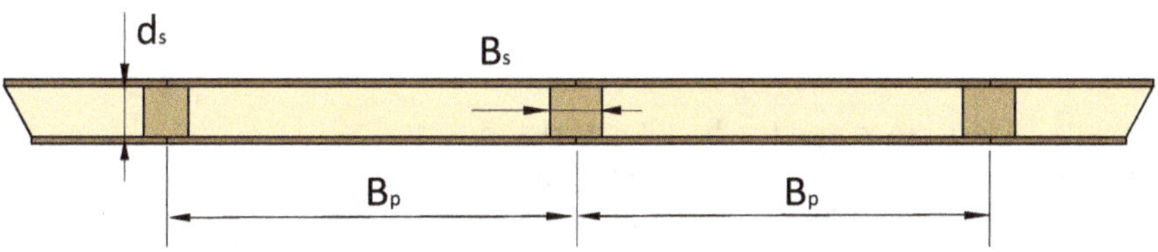

Figure 4.6.1 Wall Plan with Studs in Each Spline Joint

4.7 COMBINED AXIAL COMPRESSION AND BENDING

External walls are subject to wind loading. Checking walls for a combination of transverse bending and axial compression is based on Eurocode 5. The following equation should be satisfied:

$$\frac{F_{c,0,d}}{P_{c,0,d}} + k_m . \frac{M_{y,0,d}}{M_{r,d}} \leq 1 \text{ ......where...}$$

$F_{c,o,d}$ = Applied design compression
$P_{c,o,d}$ = Design axial compression resistance (Minimum of P_b & bearing resistance)
P_{cr} (buckling resistance), P_{cre} (eccentric buckling resistance) from previous sections
$M_{y,o,d}$ = Applied design transverse moment in panel
$M_{r,d}$ = Design moment resistance from **chapter 2** (for wind, instantaneous value).
k_m = 1.0 *(a value of 0.7 for solid rectangular sections, but is assumed 1.0, worse case).*

Combined Axial and Bending Check Example

A 2.7m high, 142mm wall with 15mm face thickness is subjected to design *ultimate* axial loading from roof, floor and wall and *ultimate* lateral wind loading as illustrated in **figure 4.7.1**. From the analysis methods described in **section 2.2.1** (bending) and **section 4.3** and **4.4** (axial), the following resistances were obtained:

$P_{c,o,d}$ = 44kN/m (Note: P_{cre} used < P_b and P_{cr})
*Note: e=24mm from **figure 4.4.3** calculation*
$M_{r,d}$ = 6.2kNm/m

Applied design (ultimate) forces and moments:

$F_{c,o,d}$ = 6 + 6 + 6 = 18kN/m (ultimate) < 28
$M_{y,0,d} = F_{lat,d}.L^2/8 + F_{w,d}.e/2$*
$= 2.25 \times 2.7^2 / 8 + 6 \times 0.071/2 = 2.1 + 0.21 = 2.31$kNm/m

**The moment due to eccentricity (50% at mid height of the moment at the head for a triangular moment diagram) is shown for completeness; low in magnitude compared with the moment due to wind load.*

Checking combined bending and axial load:

$$\frac{18}{44} + 1.0 \times \frac{2.31}{6.2} = 0.41 + 0.37 = 0.78 < 1$$

Since the axial resistance and the bending resistance is generally well within the applied moment, even with a high wind pressure of 2.25 KN/m² (equivalent to an unfactored net pressure of 1.5kN/m²), there is rarely any concern in practice. See **section 4.7.1**.

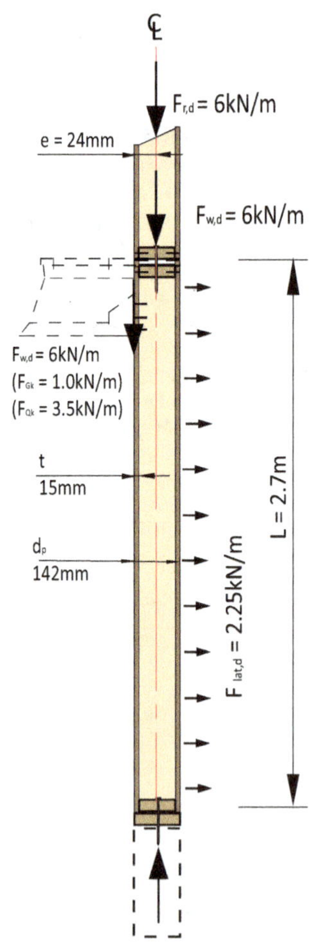

Figure 4.7.1 Combined Axial and Bending Example

4.7.1 Combined Axial Compression and Bending: Deflections

The wind panel in **figure 4.7.1** has a high wind load acting combined with eccentric loading. An assessment of the deflection from instantaneous wind loading and long term eccentric dead and live loading follows, using the method outlined in **section 2.3** and a formula for deflection under eccentric loading.

Deflection Formula Due to Out of Plane Wind Load (from Section 2.3)

$$\text{Deflection, } d_{max,w} = \frac{5.w.L^4}{384.E_f.I_f}.(1 + k_{def,f}) + \frac{w.L^2}{8.A_c.G_{c(B)}}.(1 + k_{def,c})$$

where...

E_f = Young's Modulus of skins (faces) - typically 3800 N/mm^2
assuming compression modulus = tension modulus
This should be confirmed by referring to the manufacturer's data.

I_f = Moment of inertia of faces ($I_f = (d_p^3 - d_c^3).B_p/12$)
A_c = Area of foam core (Alternatively substitute $A_{eff} = d_{eff}.B_p$ from Allen[9])
$G_{c(B)}$ = Flexural shear modulus of foam core from **section 2.3**.
$K_{def,f}$ = Load duration factor for the face
$K_{def,c}$ = Load duration factor for the core

Deflection Formula Due to Eccentricity

From BM TRADA Q-Mark: "Engineered wood products", section 9.3.2:

$$\text{Deflection, } d_{max,e} = (M_{ecc,Gk} + \psi_2.M_{ecc,Qk})\left\{\frac{L^2}{9.\sqrt{3}.E_f.I_f}.(1 + k_{def,f}) + \frac{3}{8.A_c.G_{c(B)}}.(1 + k_{def,c})\right\}$$

Where...

Eccentric moment at top of panel due to dead load, $M_{ecc,Gk} = F_{Gk}.e$
Eccentric moment at top of panel due to live load, $\psi_2.M_{ecc,Qk} = \psi_2.F_{Qk}.e$
Variable action factor for the live load (see section 2.3.2.2), $\psi_2 = 0.30$ on floors

$K_{def,f}$ = 0.0 for instantaneous loading under wind loading conditions for the faces.
$K_{def,c}$ = 0.0 for instantaneous loading under wind loading condition for the core.
See **table 2.2.1** for load duration factors.

Combined Axial and Bending Deflection Checks

Based on **figure 4.7.1**, the instantaneous deflection of the panel, at mid height, was calculated using the formulae above. For long term deflection, the wind load may be omitted from the calculation. Permanent dead load duration factors from **table 2.2.1** ($K_{def,f}$ and $K_{def,c}$) are applied with medium term live load duration factors in this instance. *The dead and live load deflection checks would have to be split into two separate calculations and added together.*

*Calculate Instantaneous Deflection due to Wind Load (from **figure 4.7.1**)*

Deflection, $d_{max,w} = \dfrac{5.w.L^4}{384.E_f.I_f} . (1 + k_{def,f}) + \dfrac{w.L^2}{8.A_c.G_{c(B)}} . (1 + k_{def,c})$

Where....

$I_f = (d_p^3 - d_c^3).B_p/12 = (142^3 - 112^3) \times 1000/12 = 122\times10^6\ mm^4/m$

$A_c = 112 \times 1000 = 112{,}000mm^2/m$

$G_{c(B)} = 2.3$ from **section 2.3**.

$K_{def,f} = 0$ and $K_{def,c} = 0$

Deflection, $d_{max,w}$ calculation (for serviceability wind load, $f_{lat,s} = 1.5/1.5 = 1.0kN.m$) :

$d_{max,w} = 5\text{x}1.5\text{x}2700^4\ \text{x}(1+0)/384/3800/\ 122\text{x}10^6 + 1.5\text{x}2700^2\ \text{x}(1+0)/8/112{,}000/2.3$
$= 2.2 + 5.3 = 7.5mm$

*Calculate Instantaneous Deflection due to Eccentricity (from **figure 4.7.1**)*

Deflection, $d_{max,e} = (M_{ecc,Gk} + \psi_2 .M_{ecc,Qk})\left\{\dfrac{L^2}{9.\sqrt{3}.\ E_f.I_f} .(1 + k_{def,f}) + \dfrac{3}{8.A_c.G_{c(B)}} .(1 + k_{def,c})\right\}$

Where, for instantaneous loading...

$M_{ecc,Gk} = F_{Gk}.e = 1 \times 0.071 = 0.071kNm/m$ (half depth eccentricity)
$\psi_2 .M_{ecc,Qk} = \psi_2.F_{Qk}.e = 0.30 \times 3.5 \times 0.071 = 0.075kNm/m$

Deflection, $d_{max,e}$ calculation:

$= (0.071+0.075)\text{x}10^6\ \text{x}\ \{2700^2\text{x}(1+0)/9/\sqrt{3}/\ 3800\ /\ 122\text{x}10^6 + 3/8/112{,}000/2.3\}$
$= 0.15 + 0.21 = 0.36mm << 7.5mm$,

Note: this would equate to 2.2mm permanent deflection using K_{def}'s from **table 2.2.1**

Total Instantaneous Deflection due to Wind and Eccentricity

Total mid height deflection of wall = $d_{max,w} + d_{max,e} = 7.5 + 0.36 = 7.9mm$ (L/342)

Note that a net wind loading suction or pressure of 1.5kN/m² is very high for a domestic wall in the UK. A more realistic loading for a ground floor wall in most circumstances, at under 100m altitude, up to 25m/s design wind speed, would be in the order of 0.8 - 1.0kN/m².

This would reduce the deflection to under 5mm (L/540), which is acceptable for all types of cladding, including masonry.

In conclusion, in most circumstances, wind loading on OSB SIP panels is not going to be of concern for combined bending and axial strength or deflection for 2-3 storey domestic properties. This condition should be considered in buildings with higher than normal floor to floor heights. An example is for a sports hall where the wall heights may approach 5-6m.

4.8 CRITICAL BUCKLING CHECKS FOR STUD WALLS

In this section, the axial compression resistance of internal stud load bearing walls will be assessed in accordance with Eurocode 5, sections 6.3.2.

Stud Bearing Resistance

For a short stud where buckling is not critical the compression resistance, P_{bs}, is as follows:

$$P_{bs} = f_{c,0,k} . k_{sys} . \frac{K_{mod}}{\gamma_M} . B_s . d_s$$ ....where.....

$f_{c,0,k}$ = Characteristic compression strength parallel to grain
K_{sys} = 1.1 (multiple stud factor) on the condition $S_s \leq 610mm$
K_{mod} = duration factor for service class
B_s = width of stud
d_s = depth of stud within wall
γ_M = material factor for grade of timber

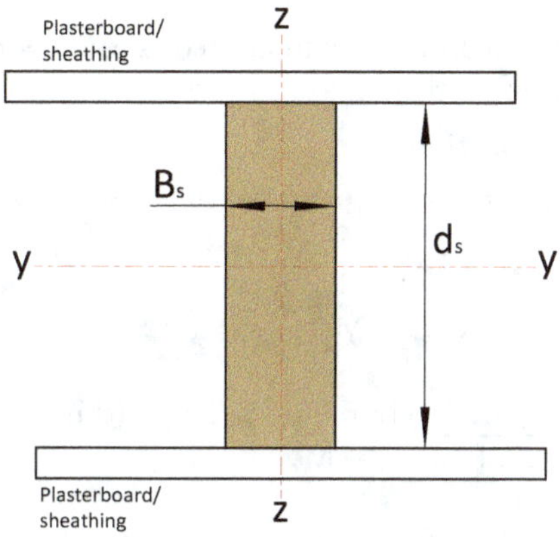

Figure 4.8.1 Stud Wall Section

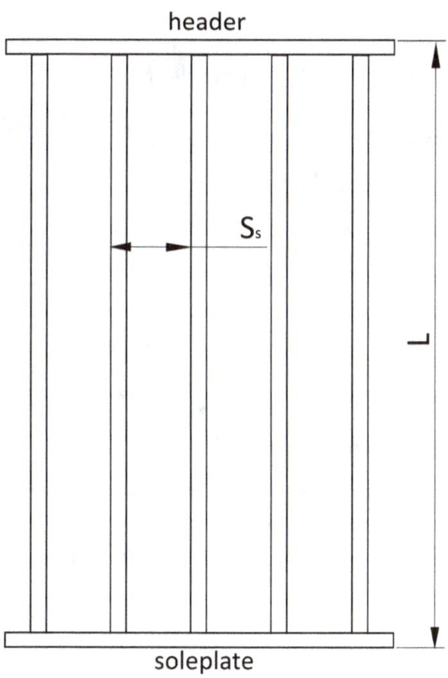

Figure 4.8.2 Stud Wall Elevation

Stud Slenderness Ratios

For slender studs, where buckling is critical, the following parameters are calculated:

For a rectangular section:

$$\text{Slenderness, } \lambda_y = \frac{L_{eff}}{r_{yy}} = \frac{L_{eff} . \sqrt{12}}{d_s}$$

$$\text{Slenderness, } \lambda_z = \frac{L_{eff}}{r_{yz}} = \frac{L_{eff} . \sqrt{12}}{B_s}$$

Relative slenderness ratios:

$$\lambda_{rel,y} = \frac{\lambda_y}{\pi} . \sqrt{\left(\frac{f_{c,0,k}}{E_{0,05}}\right)}$$

$$\lambda_{rel,z} = \frac{\lambda_z}{\pi} . \sqrt{\left(\frac{f_{c,0,k}}{E_{0,05}}\right)}$$

$E_{0,05}$ = 5th percentile modulus parallel to grain
L_{eff} = k.L where k = 1.0 or less depending on restraint condition (EC5 table 6.1)
See **figure 4.8.1** for dimensions.

Stud Buckling Combined Axial Compression and Bending Assessment

In Eurocode 5, section 6.3.1, checks for columns with combined bending includes bending about two axes: $M_{y,d}$ and $M_{z,d}$. Stud walls support joists or beams which are perpendicular to the wall. Therefore, the principal moment, My, about the Y - Y axis, is the main concern. The Z - Z moment, $M_{z,d}$, is omitted to simplify the equations below.

In accordance with Eurocode 5, section 6.3.2 and 6.3.3, the following conditions are to be satisfied:

Condition 1

$\lambda_{rel,y} > 0.3$ and ($\lambda_{rel,z}$ *any value*) and $\lambda_{rel,m} \leq 0.75$ *i.e. lateral torsional buckling does not apply*

$$\frac{F_{c,d}}{k_{c,y} \cdot P_{c,0,d}} + \frac{M_{y,d}}{M_{r,d}} \leq 1$$

$$\frac{F_{c,d}}{k_{c,z} \cdot P_{c,0,d}} + k_m \cdot \frac{M_{y,d}}{M_{r,d}} \leq 1$$ (this equation may be omitted if $\lambda_{rel,z} \leq 0.3$)

$P_{c,0,d}$ = Design compression strength parallel to grain $P_{c,0,d} = K_{sys} \cdot \dfrac{f_{c,0,k} \cdot K_{mod} \cdot B_s \cdot d_s}{\gamma_M}$

See previous page.

$F_{c,d}$ = Design applied axial compression load

$M_{y,d}$ = Design applied moment at mid height

k_m = 0.7 for rectangular sections (1.0 for other cross sections)

$M_{r,d}$ = Design resistance moment for the condition k_{crit} = 1 (see overleaf)

Buckling factors, $k_{c,y}$ and $k_{c,z}$ are obtained from the equations below:

$$k_{c,y} = \frac{1}{k_y + \sqrt{(k_y^2 - \lambda_{rel,y}^2)}}$$

$$k_y = 0.5\,(1 + \beta_c\,(\lambda_{rel,y} - 0.3) + \lambda_{rel,y}^2)$$

$$k_{c,z} = \frac{1}{k_z + \sqrt{(k_z^2 - \lambda_{rel,z}^2)}}$$

$$k_z = 0.5\,(1 + \beta_c\,(\lambda_{rel,z} - 0.3) + \lambda_{rel,z}^2)$$

Straightness factor, β_c = 0.2 (solid timber) and 0.1 (glue laminated timber or glulam)

Compression resistance, $P_{c,0,d} = k_{c,y} \cdot P_{bs}$ *or* $k_{c,z} \cdot P_{bs}$ *...whichever is the minimum value.*

Calculation of $\lambda_{rel,m}$

$$\lambda_{rel,m} = \sqrt{\left(\frac{f_{m,k}}{\sigma_{m,crit}}\right)}$$ ...where, for a rectangular section:

$$\sigma_{m,crit} = \frac{0.78\,B_s^2}{d_s \cdot L_{eff}}$$

$f_{m,k}$ = Characteristic bending strength, *noting that...*

$f_{m,d}$ = Design bending strength = $f_{sys} \cdot \dfrac{K_{mod} \cdot f_{m,k}}{\gamma_M}$

Condition 2

$\lambda_{rel,y} > 0.3$ and/or $\lambda_{rel,z} > 0.3$ and $\lambda_{rel,m} \geq 0.75$ *i.e. lateral torsional buckling applies*

$$\frac{F_{c,d}}{k_{c,z} \cdot P_{c,0,d}} + \left(\frac{M_{y,d}}{M_{r,d}}\right)^2 \leq 1$$

Since moments about the Z - Z axis are omitted, the design resistance moment, $M_{r,d}$ is obtained from the equation:

$$M_{r,d} = k_{crit} \cdot f_{m,d} \cdot Z_y = k_{crit} \cdot f_{m,d} \cdot \frac{B_s \cdot d_s^2}{6}$$

k_{crit} is a factor which takes into account the reduced bending strength due to lateral buckling, which may be obtained from:

$\lambda_{rel,m} \leq 0.75$	$k_{crit} = 1$
$0.75 < \lambda_{rel,m} \leq 1.4$	$k_{crit} = 1.56 - 0.75\ \lambda_{rel,m}$
$\lambda_{rel,m} > 1.4$	$k_{crit} = 1 / \lambda_{rel,m}^2$

Condition 3

$\lambda_{rel,y} < 0.3$ and $\lambda_{rel,z} < 0.3$ and $\lambda_{rel,m} \geq 0.75$ *i.e.lateral torsional buckling applies*

Check all three equations for both *condition 1* and *condition 2*

Condition 4

$\lambda_{rel,y} > 0.3$ and $\lambda_{rel,z} < 0.3$ and $\lambda_{rel,m} < 0.75$ *i.e.lateral torsional buckling does not apply*

$$\left(\frac{F_{c,d}}{P_{c,0,d}}\right)^2 + \frac{M_{y,d}}{M_{r,d}} \leq 1 \quad \textit{noting that } k_{crit} = 1$$

In practice, this condition is unlikely to occur.

Note on Sheathed Studs

Sheathed walls are generally constructed from OSB or plywood sheets, as discussed in **chapter 3** on racking. Plasterboard may be considered to act as sheathing if sufficient fixings to the studs are applied, with improved or threaded nails or screws fixed to the studs as centres not less than 150mm; mostly ignored unless the specifier is confident that this will be applied on site.

If a wall is considered to be sheathed, the minor Z - Z axis in **figure 4.8.1** is not critical to the design since the sheathing will restrain studs from buckling. In this case the relative slenderness will only be relevant to the major axis, Y - Y.

In the following examples, which consider only axial loading, which in most cases is sufficient for checking studs, unsheathed and sheathed walls will be reviewed, along with a wall restrained at half height by noggins.

Example: Stud Buckling Under Axial Compression with no Sheathing

From **figure 4.8.2** 38x90 studs (B_s = 38mm, d_s = 90mm) are 2.7m high supporting joists and fixed to a sleeper wall below. The studs are grade C16 timber. The design loading is 17.5kN/m from the roof, the wall above (and half wall below) and joists. There is no eccentric loading or lateral loading to consider.

C16 timber: $f_{m,k}$ = 16N/mm²; $f_{c,0,k}$ = 17N/mm²; $E_{0,05}$ = 5400N/mm² (from BS EN 338)

Effective length coefficient, k = 0.90, assuming lateral restraint from walls and floor, reducing the buckling length of the column.

$L_{eff} = 0.9 \times 2.7 = 2.43m \approx 2.4m$

Critical slenderness is in the Z - Z axis

$$\text{Slenderness, } \lambda_z = \frac{L_{eff}}{r_{yz}} = \frac{L_{eff} . \sqrt{12}}{B_s} = 2400 \times 3.46 / 38 = 219$$

Relative slenderness is calculated from:

$$\lambda_{rel,z} = \frac{\lambda_z}{\pi} . \sqrt{\left(\frac{f_{c,0,k}}{E_{0,05}}\right)} = 219/3.142 \times \sqrt{(17/5400)} = 3.9$$

$k_z = 0.5\ (1 + \beta_c\ (\lambda_{rel,z} - 0.3) + \lambda_{rel,z}^2)$, where $\beta_c = 0.2$, for solid timber

$k_z = 0.5\ (1 + 0.2 \times (3.9 - 0.3)+3.9^2) = 8.5$

$$k_{c,z} = \frac{1}{k_z + \sqrt{(k_z^2 - \lambda_{rel,z}^2)}} = 1 / \{8.5 + \sqrt{(8.5^2 - 3.9^2)}\} = 0.06$$

Axial compression resistance, long term ($k_{mod} = 0.8$, $\gamma_M = 1.3$), $P_{c,0,d}$ is found:

$$P_{bs} = \frac{f_{c,0,k}.k_{sys}.K_{mod}.\ B_s.d_s}{\gamma_M} = 17 \times 1.1 \times 0.8 \times 38 \times 90 / 1.3/1000 = 39.3kN$$

$P_{c,0,d} = k_{c,z} . P_{bs} = 0.06 \times 39.3 = 2.4kN$

Resistance of wall:

Studs at 600mm centres, $P_{w,o,d(600)} = 2.4/0.6 = 4kN/m << 17.5$

Studs at 400mm centres, $P_{w,o,d(400)} = 2.4/0.4 = 6kN/m << 17.5$

Wall is insufficient without sheathing or additional restraint.

Example: Stud Buckling Under Axial Compression with Sheathing

From the above example the C16 studs are checked with sheathing restraining the Z - Z axis.

Critical slenderness is in the Y - Y axis

$$\text{Slenderness, } \lambda_y = \frac{L_{eff}}{r_{yy}} = \frac{L_{eff} . \sqrt{12}}{d_s} = 2400 \times 3.46 / 90 = 92$$

Relative slenderness is calculated from:

$$\lambda_{rel,y} = \frac{\lambda_y}{\pi} \sqrt{\left(\frac{f_{c,0,k}}{E_{0,05}}\right)} = 92/3.142 \times \sqrt{(17/5400)} = 1.65$$

$k_y = 0.5\ (1 + \beta_c\ (\lambda_{rel,y} - 0.3) + \lambda_{rel,y}^2)$, where $\beta_c = 0.2$
$k_y = 0.5\ (1 + 0.2 \times (1.65 - 0.3) + 1.65^2) = 2.0$

$$k_{c,y} = \frac{1}{k_y + \sqrt{(k_y^2 - \lambda_{rel,y}^2)}} = 1 / \{2 + \sqrt{(2^2 - 1.65^2)}\} = 0.32$$

Axial compression resistance, long term ($k_{mod} = 0.8$, $\Upsilon_M = 1.3$), $P_{c,0,d}$ is found:

$$P_{bs} = f_{c,0,k}.k_{sys}.\frac{K_{mod}}{\Upsilon_M}.B_s.d_s = 17 \times 1.1 \times 0.8 \times 38 \times 90 / 1.3/1000 = 39.3\text{kN}$$

$P_{c,0,d} = k_{c,z}\ .\ P_{bs} = 0.32 \times 39.3 = 12.6$kN (>> 2.4kN *unsheathed value from previous page).*

Resistance of wall:
Studs at 600mm centres, $P_{w,o,d(600)} = 12.6/0.6 = 21.0$kN/m > 17.5
Studs at 400mm centres, $P_{w,o,d(400)} = 12.6/0.4 = 31.5$kN/m > 17.5

Wall is sufficient with sheathing.

Example: Stud Buckling Under Axial Compression with Half Height Restraint

The wall is checked to find out whether a half height restraint from noggins at half height with raking studs for restraining buckling is adequate, with no sheathing.

$L_{eff} = 2.7/2 = 1.35$m (in Z - Z axis, but remains 2.4 for the Y - Y axis).

Slenderness, $\lambda_z = \dfrac{L_{eff}}{r_{yz}} = \dfrac{L_{eff}.\sqrt{12}}{B_s} = 1350 \times 3.46 / 38 = 123 > 90$ - *Z - Z buckling critical*

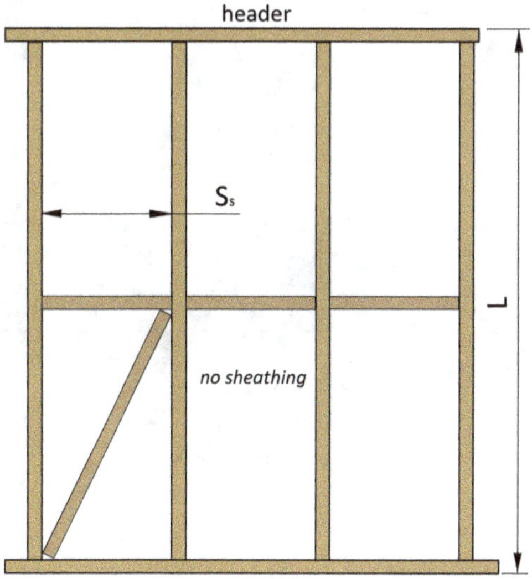

Figure 4.8.3 Stud Wall Elevation with Half Height Restraint

$\lambda_{rel,z} = 123/3.142 \times \sqrt{(17/5400)} = 2.2$

$k_z = 0.5\ (1 + 0.2 \times (2.2-0.3) + 2.2^2) = 3.1$

$k_{c,z} = 1 / \{3.1 + \sqrt{(3.1^2 - 2.2^2)}\} = 0.19$

$P_{c,0,d} = 0.19 \times 39.3 = 7.4$kN

Resistance of wall:
$P_{w,o,d(600)} = 7.4/0.6 = 12.3$kN/m < 17.5
$P_{w,o,d(400)} = 7.4/0.4 = 18.5$kN/m > 17.5

The wall is adequate with studs at 400mm centres unsheathed.

Eccentric Loading

If there is eccentric loading or lateral loading to consider, the calculations would have to include for combined bending and compression. Formulae may be derived from Eurocode 5 section 6.3.3.

4.9 WALLS SUPPORTING PRECAST FLOORS

Hotels have been built with integrated SIPs, internal and external walls and precast floors. With precast panels spanning up to circa 6m onto double party walls, providing acoustic separation, the floor loadings can be high compared with timber floor construction. Without knowledge of the actual floor to wall details, a solution was proposed in a previous project, as illustrated in **figure 4.9.1** below, for a three storey office construction with 5m span precast floors with structural screed.

Progressive collapse was a concern with the floors, entailing the removal of sections of walls, which was checked in a manner outlined in **section 3.6.2**. The precast floor with top reinforcement and screed was shown to maintain the integrity of floors supporting panels above, spanning circa 2.4m over the walls below. For greater widths, glulam header beams were considered under the precast floor to assist. Twisted straps in every joint of the 1.2m precast panel were shown cast into the screed for robust wall ties. An additional 150mm wide header was added to spread the load.

The design loading for a 5m span, including a joisted roof, for the ground floor wall was 32kN/m dead load and 8kN/m live load. This totalled 54kN/m ultimate loading. In **section 4.3** a 3m high 125mm SIP wall was shown to have a design resistance of 43kN/m (<54kN/m). For the proposed 150mm wall with 100x125 studs in every 1200mm joint this was shown to be adequate, assuming no eccentricity of load.

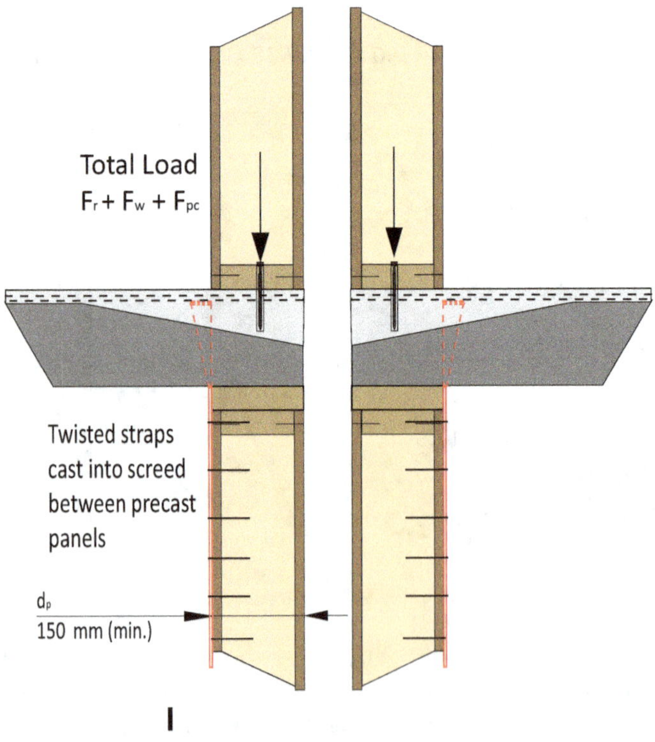

Figure 4.9.1 SIP Party Wall with Precast Concrete Floor

4.10 MgO (MAGNESIUM OXIDE) PANEL WALLS

Wall panels have been widely adopted in SIPs construction, including in garden offices. They are not well suited to roof construction owing to their relative brittleness in transverse bending. MgO may be reinforced with glass fibres to enhance its tensile and flexural strength to circa 22N/mm². Typical indicative standard board properties for unreinforced panels are shown below:

Density 1000 kg/m³
$f_{t,0,k(MgO)}$ = Tensile Strength = 2.6 N/mm²
Flexural tension strength 3.6 N/mm²
MgO Compression Strength: 11 N/mm²
Flexural modulus 6400 N/mm²
Ef = Young's Modulus 4650 N/mm²
(assuming $E_{f(0.05)}$ = 0.7 x 4650 = 3255N/mm²)
Shear modulus circa 2000 N/mm²

Note that creep properties are not critical to wall panels. For the example in **figure 4.10.1**., based on the methods in **section 4.3**, compression and bending resistance, ignoring eccentricity, are calculated ($G_{c(B)}$ =2.3N/mm²):

$$P_{cr} = \frac{P_E}{1 + \frac{P_E}{A_{eff}.G_{c(B)}}} \quad \text{... where } P_E = \frac{\pi^2.D}{L_{eff}^2}$$

and flexural rigidity $D = \frac{E_f.B_p.t.d_{eff}^2}{2}$

D = 3255x1000x12x(150-12)² / 2 = 372x10⁹Nmm²
P_E = 3.142²x372x10⁹/2700²/1000 = 504kN/m
P_{cr} = 504/(1+504x10³/(150-12)/1000/2.3) = 195kN/m

$$P_{cr,d} = \frac{P_{cr}.\ k_{mod(MgO)}}{\gamma_M} = 195 \times 0.40/1.3 = 60\text{kN/m} > 25$$

Bending strength instantaneous from **section 2.3.5** (γ_m, particle board = 1.3):

$$M_{d,t} = \frac{f_{t,o,k}\ .\ B_p.\ t_2.\ K_{mod}.\ d_{eff}}{\gamma_m}$$ (from **section 2.2.1**)

$M_{d,t}$ = 2.6x1000x12x1.0x(150-12)/1.3/1x10⁶
= 3.3kNm/m

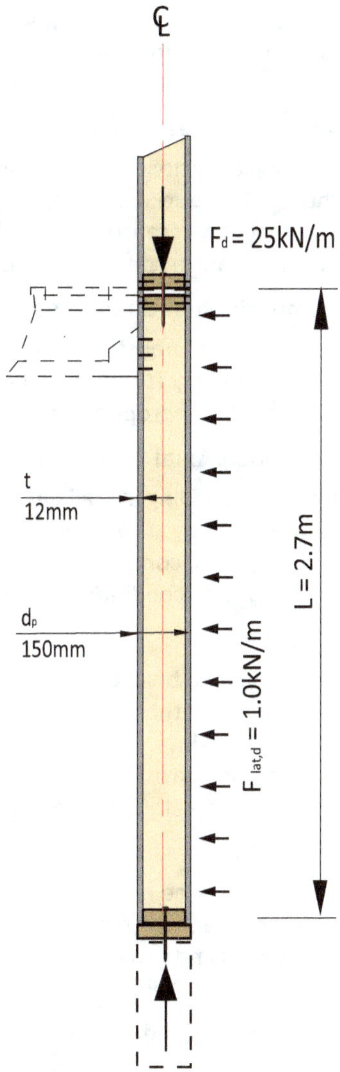

Figure 4.10.1 SIP MgO Wall Example

$M_u = wL^2/8$ = 1 x 2.7² / 8 = 0.91kNm/m < 3.3

From **section 4.6**, combined bending and compression check:
0.91/3.3 + 25/60 = 0.69 < 1 *...the wall is adequate.*

Chapter 4

Quick Note Summary

Axial Compression

There are three principle modes of failure in compression: bulging of the faces, peeling at header and soleplate and buckling. Walls are checked for bearing in the faces based on the face compression strength and buckling critical load. For most panels in a wall at a 2.5-3m storey height, buckling governs the design.

Axial Compression with Eccentric Loading

Buckling capacity is reduced when loading is eccentric. A secant formula for assessing the reduced buckling capacity is appropriate for wall panels. Joists can provide resistance to eccentric loading by means of their wall connection, but ignoring this ensures a more conservative assessment.

Comparison of Calculated and Tested Axial Compression Values

Table 4.5.1 demonstrates, that with the appropriate duration factor, k_{mod}, and material factor, γ_M, the calculation method for buckling by Allen[2] and eccentricity presented is appropriate for design.

Combined Axial Compression and Bending Capacity

Axial and bending in wall panels occurs in situations such as external walls subjected to wind loading. The combined effect must be checked. For most construction types involving SIPs, compression in OSB faces and instantaneous bending in walls is not a critical design condition.

Stud Walls

Stud walls can be checked for axial compression and bending capacities based on calculation methods from Eurocode 5. The compression capacity may be enhanced by sheathing or intermediate restraint by noggins with raking struts, such that the minor axis of studs is restrained. Studs, as multiple elements, may have enhanced capacity, which is calculated using a k_{sys} factor of 1.1.

Precast Floors

SIP walls may be employed to support precast floors. Additional restraint straps at the header, which may be combined with a deep header, are needed to ensure robustness and provide resistance to progressive collapse. Timber splines in panels are recommended.

Magnesium Oxide (MgO) SIPs Panels

MgO boards are commonly used for SIPs. In countries where termites are an issue, such as in Africa, MgO SIPs are frequently adopted as an alternative to OSB-faced panels.

MgO faces are similar to OSB in relation to compressive strength and stiffness but have a relatively low tensile strength, which means that for roofs and floors, OSB SIPs are more appropriate. MgO compression and bending resistance may be enhanced by timber studs, in walls and lintels.

Chapter 5
SIPs as a Beam

SIP wall as a deep beam. SIP panel as a lintel. First principles in flexure. Lintel detail. Multi panel SIP walls as deep beams. The issues of shear transfer at the OSB/timber interface. MgO faced lintels.

5.1 SIPS WALL AND LINTEL PANELS DESIGNED AS BEAMS

SIPs panels over openings with OSB faces, and timber trimming to the top and bottom are essentially the same structurally as engineered joists with OSB webs with graded timber flanges.

Figure 5.1.1 shows a first floor, wall that is designed to span over a large opening.

At first floor the deeper lintel section is designed to support the wall during construction. A continuous header sits on the beam, acting as a soleplate for the walls above and support for the joist hangers behind. The total span is 3.7m (3 panels width). For permanent loads from the roof and wall, the floor panels and header above, which are all nailed and adhesive bonded together, may be designed as a deep beam, with compression in the eaves headbinder, tension in the lintel at first floor and in-plane shear carried by the wall panels.

Figure 5.1.2 overleaf shows a wall with large openings on the ground floor and first floor. The lintels are designed as composite beams to carry the first floor and roof, thereby reducing the size of the lintel had it been designed without assuming the structural assistance of the SIP panels. Sections at the first floor and at the eaves are illustrated in **figures 5.1.3** and **5.1.4,** respectively.

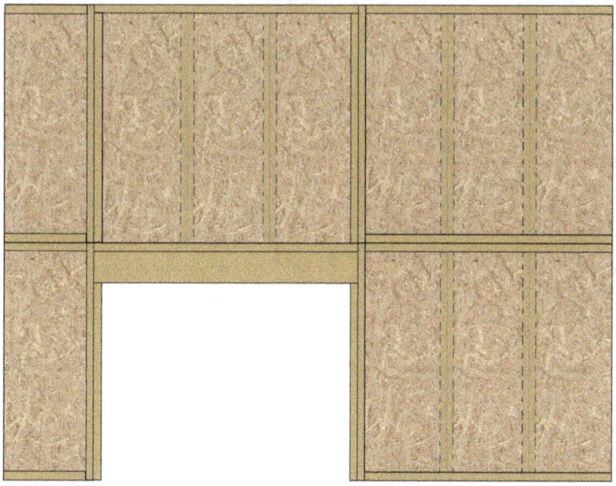

Figure 5.1.1 Sips Panels Spanning over Opening

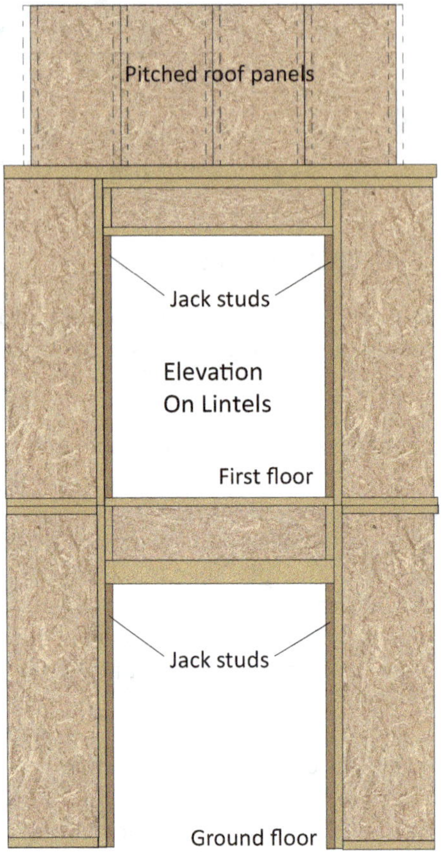

Figure 5.1.2 SIPs Lintels: First Floor and Eaves

For the wall shown in **figure 5.1.1** and the lintels in **figure 5.1.2**, examples follow, which demonstrate the composite action of SIPs and timber lintels.

In **section 5.2**, a method of assessing the structural load bearing capacity of composite SIPS and timber assemblies is outlined.

Efficacy of SIPs Composite Beams

Whilst this is a useful application of SIPs construction, which has the potential to reduce or eliminate large timber lintel sections, successful installation depends on attention to assembly and erection.

The recessed joints on all sides of a panel must be adhesively bonded, as will be demonstrated in the examples. The SIP beams should sit on a *jack stud* (outermost stud), which is then fixed to the inner *king stud,* ensuring a robust transfer of shear to supporting posts. See **figure 8.3.16** Off site assembly for lintel or wall components avoids the greater risk of site assembled inadequate joints, adhesive bonded in a less controlled environment.

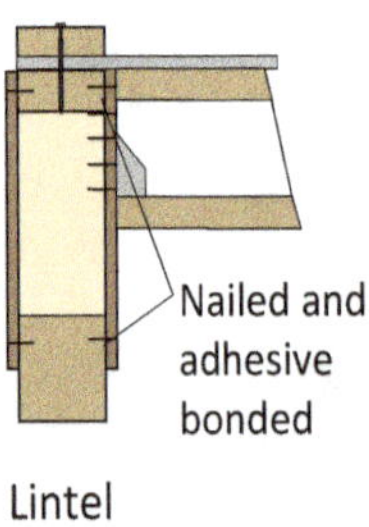

Figure 5.1.3 Section at First Floor

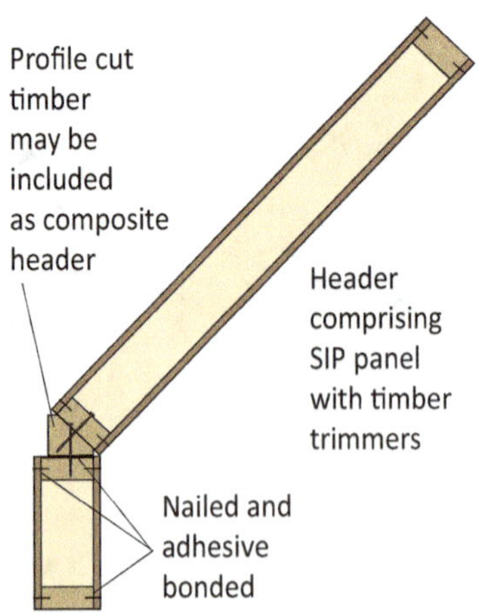

Figure 5.1.4 Section at Eaves

Employing the greater economy of SIPs beams is better suited to projects with repetition of openings, typically large houses, offices or schools.

The analytical finite element may be used to demonstrate walls. An example is presented in **section 3.6.2** for a progressive collapse condition.

5.2 SIPS WALL AND LINTEL PANELS DESIGNED AS BEAMS WITH EXAMPLES

The wall shown below in **figure 5.2.1** on the first floor of a house supports a floor and roof. It is reviewed as a deep beam with a header and first floor lintel as flanges and SIP panels as the shear resisting web.

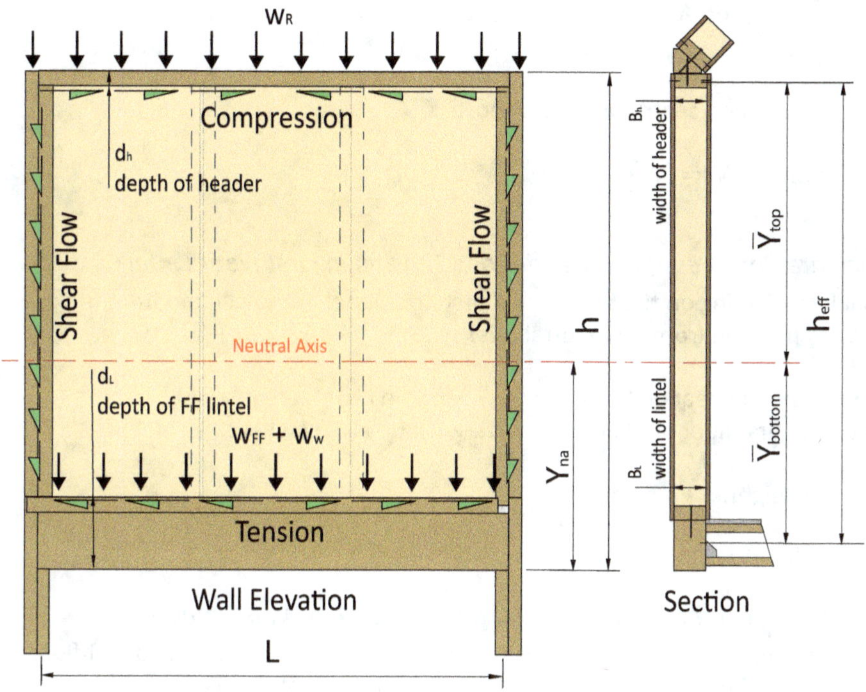

Figure 5.2.1 SIP Wall as a Beam Internal Forces

Internal Forces

On the assumption that the primary action in the header and first floor lintel are axial, based on the depth of the beam limiting deflections, and hence flexural deformation, the internal forces may be calculated from the following formulae.

Calculate Neutral Axis Position, Y_{na}

Neutral axis position from the base (soffit) of first floor lintel, taking moments of area of header and lintel:

$$Y_{na} = \frac{A_L.d_L/2 + A_h.(h-d_h/2)}{(A_L + A_h)} \quad \text{...where...}$$

A_L = Area of first floor lintel = $B_L.d_L$ (B_L = breadth/width and d_L = depth of lintel)
A_h = Area of header = $B_h.d_h$ (B_h = breadth/width and d_h = depth of header)
h = overall height of panel including lintel and header depth

Calculate Design Bending Moment and Support Shear

In **figure 5.2.1,** the loads acting on the wall are based on the first floor joist loading (W_{FF}), roof loading (W_R) and wall self weight (W_w). The total ultimate load acting on the wall (W_d) is a summation of ultimate (design) dead and live loads from these three components.

Design Moment, $M_d = \dfrac{W_d.L^2}{8}$ ...where...

$W_d = W_R + W_{FF} + W_w$

L = span between centre of support posts

Design Shear, $V_d = \dfrac{W_d.L}{2}$

Calculate Compression in Header and Tension in First Floor Lintel

Axial forces acting on the header and first floor lintel are calculated from their centroid position from the neutral axis:

Compression in header, $F_h = M_d / \bar{Y}_{top}$ ...(in kN)

Tension in first floor lintel, $F_L = M_d / \bar{Y}_{bottom}$ ...(in kN) ...where...

Effective height, $h_{eff} = h - d_L/2 - d_h/2$

$\bar{Y}_{top} = h - Y_{na} - d_h/2$

$\bar{Y}_{bottom} = Y_{na} - d_L/2$

Calculate Maximum Shear Flow (Rolling Shear) in Wall Panels

Shear within the SIP panels is transferred into the flanges (header and first floor lintel) by means of the bonded shear interface on both sides of the flange.

Shear flow at position of header, $V_{ph} = \dfrac{V_d.\ A_h.\ \bar{Y}_{top}}{2.I_p}$ *per face*..where...

Similarly, shear flow (rolling shear) at position of lintel (V_{PL}) is found by substituting $\bar{Y}_{bottom}$ into the above equation.

V_d, A_h and $\bar{Y}_{top}$ are calculated from the previous equations.

Moment of inertia of panel beam, $I_p = A_L.\ \bar{Y}_{bottom}^2 + A_h.\ \bar{Y}_{top}^2$

Note that Ip is based on the flanges for the idealised beam, and conservatively ignores the contribution of the OSB faces.

Calculate Wall Panel Deflection

This is found from combined bending and shear deflection, similar to the equation in **section 2.3**.

Deflection, $d_{max} = \dfrac{5.w.L^4}{384.E_t.I_p}.(1 + k_{def,t}) + \dfrac{w.L^2}{8.A_f.G_f}.(1 + k_{def,f})$ ...where...

E_t = Modulus of timber, Af = area of both faces and Gf = shear modulus of the faces

5.2.1 SIPs Wall as Beam Example

In **figure 5.2.2**, the SIP wall (172mm with 15mm OSB faces) spans 3.7m. It supports a floor and roof with unfactored loading shown. The first floor lintel is a 140x240mm glulam, grade GL24h, which carries the first floor loads. The header is a 140x38mm, grade C24 section with a fillet of average 240x45mm section, glued and screwed to the header. The soleplate is a 140x38mm grade C24 section. The wall is to support its own weight and roof load. From the formulae in **section 5.2** the internal forces are calculated below.

Calculate Neutral Axis Position, Y_{na}

From the base of the lintel, the neutral axis position is calculated:

$$Y_{na} = \frac{A_{sp}.d_{sp}/2 + A_h.(h-d_h/2)}{(A_{sp} + A_h)} \quad ...\text{where}...$$

A_L = Area of soleplate = $B_{sp}.d_{sp}$ = 140x38 = 5320mm²
A_h = Area of header = $B_h.d_h$ = 140x(38+45) = 11,620mm²
h = 2.5m

$$Y_{na} = \frac{5320\text{x}38/2 + 11{,}620 \text{ x } (2500 - (38+45)/2)}{(5320 + 11{,}620)} = 1692\text{mm}$$

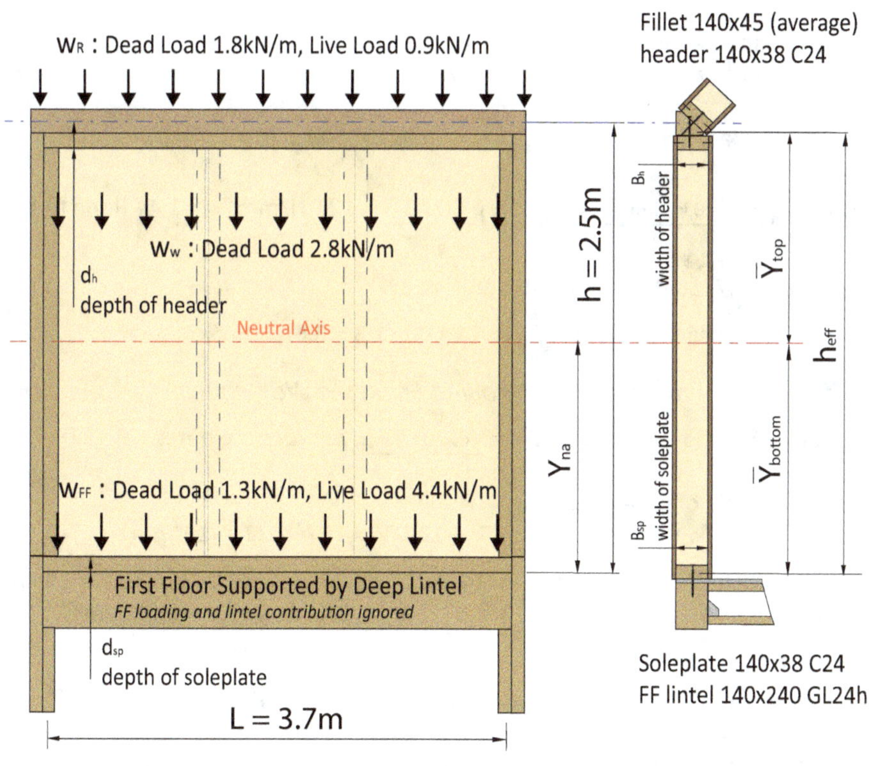

Figure 5.2.2 172mm SIP Wall as a Beam Example

Calculate Design Bending Moment and Support Shear

In **figure 5.2.2,** the loads acting on the wall are based on: roof loading (w_R) and wall self weight (w_w); *omitting first floor joist loading* (w_{FF}), *which is carried by the lintel.*

Total dead load, $W_{dead} = W_{Rd} + W_{wd} = 1.8 + 2.8 = 4.6$kN/m
Total live load, $W_{live} = W_{Rl} + W_{wl} = 0.9 + (0) = 0.9$kN/m

Design load, $W_d = 4.6 \times 1.35 + 0.9 \times 1.5 = 7.6$kN/m

Design Moment, $M_d = W.L^2/8 = 7.6 \times 3.7^2 / 8 = 12.9$kNm
Design Shear, $V_d = W_d.L/2 = 7.6 \times 3.7 / 2 = 14.1$kN

Calculate Compression in Header and Tension in Soleplate

Axial forces acting on the header and first floor lintel are calculated from their centroid position from the neutral axis:

Compression in header, $F_h = M_d / \bar{Y}_{top}$ ...(in kN)
Tension in soleplate, $F_{sp} = M_d / \bar{Y}_{bottom}$ ...(in kN) ...where...

Effective height, $h_{eff} = h - (d_{sp} + d_h)/2 = 2500 - (38+38+45)/2 = 2439$mm (2.43m)
$\bar{Y}_{top} = h - Y_{na} = 2500 - \{1692 - (38 + 45)/2\} = 767$mm (0.767m)
$\bar{Y}_{bottom} = Y_{na} - d_{sp}/2 = 1692 - 38/2 = 1673$mm (1.673m)

Compression in header, $F_h = 12.9/0.767 = 16.8$kN
Tension in soleplate, $F_{sp} = 12.9/1.673 = 7.7$kN

Compressive stress in header, $f_{ch} = F_h/A_h = 16.8\text{x}10^3 / 11{,}620 = 1.44\text{N/mm}^2$

Capacity, $f_{c,0,d} = \dfrac{f_{c,0,k} \,.\, k_{mod(osb)}}{\gamma_M} = 21 \times 0.80 / 1.3 = 12.9\text{N/mm}^2 >> 1.44$ (long term)

Similarly, $f_{tsp} = 7.7\text{x}10^3/5320 = 1.44\text{N/mm}^2 << 8.6$

(where design tensile resistance, $f_{t,0,d} = 14 \times 0.8/1.3 = 8.6\text{N/mm}^2$)

Calculate Maximum Shear Flow (Rolling Shear) in Wall Panels

Shear flow at position of header, $V_{ph} = \dfrac{V_d.\ A_h.\ \bar{Y}_{top}}{2.\ I_p}$ *per face* ...where...

Moment of inertia of panel beam, I_p:
$I_p = A_{sp}.\ \bar{Y}_{bottom}^2 + A_h.\ \bar{Y}_{top}^2 = 5320 \times 1673^2 + 11{,}620 \times 767^2 = 21.7\text{x}10^9\text{mm}^4$

Shear flow at position of header, $V_{ph} = 14.1\text{x}10^3 \times 11{,}620 \times 767/2/21.7\text{x}10^9 = 2.9$kN/m
Shear flow at soleplate, $V_{sp} = 14.1\text{x}10^3 \times 5320 \times 1673/2/21.7\text{x}10^9 = 2.9$kN/m

Note: shear flow and stresses should be equal in soleplate and header.

Chapter 3, **section 3.2.1**, summarises racking strengths for Kingspan panels of 4.4kN/m (≈ 6.4 design) which is greater than the calculated shear flow above. This is based on defined deflection limits under load. Nails are impractical to resist this loading; the bond stress between the face and timber (14.1/38 ≈ 0.4N/mm²) must be resisted by the adhesive or sufficient screws.

Calculate Permanent Wall Panel Dead Load Deflection

From the equation presented previously:

Deflection, $d_{max} = \dfrac{5.w.L^4}{384.E_t.I_p} . (1 + k_{def,t}) + \dfrac{w.L^2}{8.A_f.G_f} . (1 + k_{def,f})$ ...where...

$E_{t,C24}$ = 11,000N/mm² (GL24h assumed the same)
$k_{def,t}$ = 0.8 and $k_{def,f}$ = 2.3
Area of OSB faces, Af = 15 x 2 x 2500 = 75,000mm²
Shear modulus of OSB faces, $G_f \approx$ 1700N/mm²

Deflection, $d_{max(dead)}$ *using values obtained from the previous page*

= 5x4.6x3700⁴x(1+0.8)/384/11,000/21.7x10⁹ + 4.6x3700²x(1+2.3)/8/75,000/1700
= 0.08 + 0.20 = 0.28mm ...*very low. Live load deflection can be found using this formula.*

Note that shear deflection (0.20) is much greater than bending deflection (0.08), as would be expected for a deep beam.

Figure 5.2.3 shows the deflections from a finite element model of the SIPs wall including floor load. Maximum dead load deflection is 0.51mm.

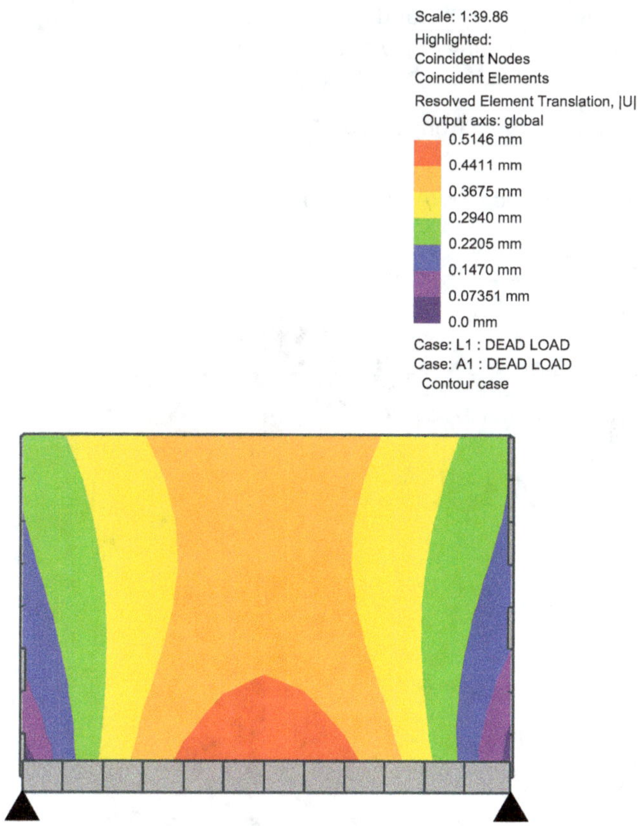

Figure 5.2.3 172mm SIP Wall Finite Element Deflection

SIPs Wall Conclusion

The SIPs wall calculation method is a reasonable representation of composite deep beam action.

A calculation for an appropriate lintel at the first floor - ignoring any stiffening from the SIPs panels over - to Eurocode 5 obtained a glulam beam size of 140x240mm deep, with a total deflection under dead and live load of 11.8mm (span/314 ratio). A first floor lintel designed to carry floor, wall and roof loads would be 140x320mm. This is an indication that *assuming deep beam properties has limited application in practice*, with an insignificant saving in the lintel size.

Notwithstanding its limitations, these calculations show that that the SIP panels will stiffen the structure significantly.

Progressive Collapse Assessment

The method outlined for the wall could be undertaken for checking progressive collapse conditions assuming wall removal under a load bearing upper wall, applying reduced load factors. Alternatively, a finite element model could be applied. See **section 3.6.2** showing output for a wall under progressive collapse conditions.

5.2.2 SIPs Composite Lintel Example

In **figure 5.2.4,** a 400mm deep lintel with standard timber splines and horizontally laid SIP panel carries a flat roof spanning 4m with dead load 0.8kN/m² (4/2 x 0.8 = 1.6kN/m) and live load 0.75kN/m² (4/2 x 0.75 = 1.5kN/m). The span is 1.2m.

Calculations follow based on formulae from **section 5.2**.

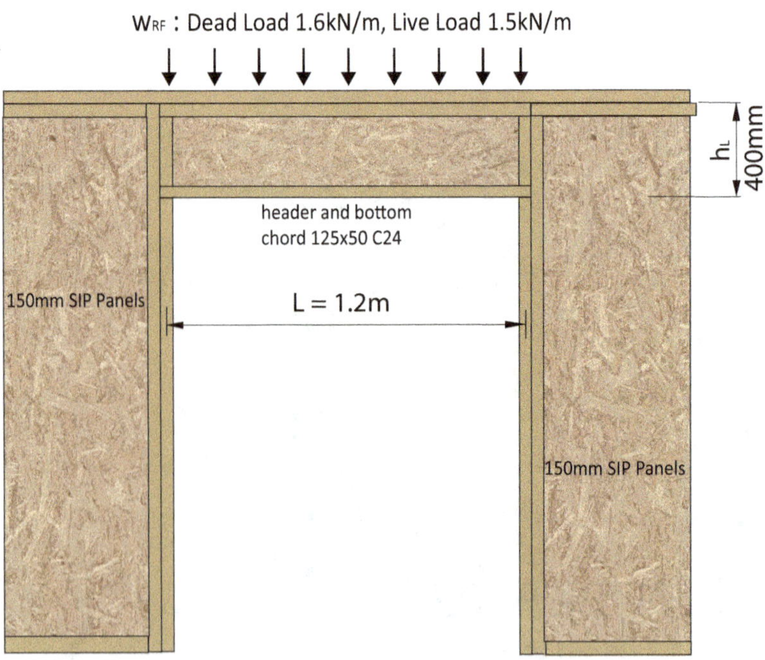

Figure 5.2.4 150mm SIP Lintel Panel (2m span at First Floor)

Calculate Design Bending Moment and Support Shear

In **figure 5.2.4,** the loads acting on the wall are based on the roof loading (w_{RF}). Header and bottom plate are 125x50mm deep (C24) (11mm OSB faces in SBS panel)

Dead load, $W_{GK} = W_{RF(dead)} = 1.6kN/m$
Live load, $W_{QK} = W_{RF(live)} = 1.5\ kN/m$

Design load, $W_d = 1.6 \times 1.35 + 1.5 \times 1.5 = 4.4kN/m$
Design Moment, $M_d = W.L^2/8 = 4.4 \times 1.2^2 / 8 = 0.8kNm$
Design Shear, $V_d = W_d.L/2 = 4.4 \times 1.2 / 2 = 2.6kN$

Calculate Compression in Header and Tension in Bottom Plate

Axial forces acting on the header and first floor lintel are calculated from their centroid position from the neutral axis, *which is central.*

Compression in header, $F_h = M_d / \bar{Y}_{top}$ ... $\bar{Y}_{top} = (400\text{-}50)/2 = 175mm$
Tension in soleplate, $F_{bp} = M_d / \bar{Y}_{bottom} = F_h$ ...*symmetrical section*

Tension in bottom plate, $F_{bp} = 0.8/0.175 = 4.6kN$ (= F_h *compression*)
Tensile stress in bottom plate, $f_{tbp} = F_{bp}/A_{bp} = 4.6x10^3 /125 / 50 = 0.74N/mm^2 < 8.6$

Calculate Maximum Shear Flow (Rolling Shear) in Lintel - at OSB Interface

Shear flow at position of header, $V_{ph} = \dfrac{V_d.\ A_h.\ \bar{Y}_{top}}{2.I_p}$ ...where...

Moment of inertia of panel beam, I_p:
$I_p = A_{sp}.\ \bar{Y}_{bottom}^2 + A_h.\ \bar{Y}_{top}^2 = 125 \times 50 \times (200\text{-}25)^2 \times 2 = 383x10^6mm^4$

Shear flow at position of header, $V_{ph} = 2.6x10^3 \times 125x\ 50 \times 175/2/\ 383x10^6 = 3.7kN/m$

Equivalent unfactored rolling shear of 2.6kN/m which is equivalent to a racking resistance of 5.4kN/m for an SBS panel. As noted in **section 3.2.7** *actual failure load on SBS panels was circa 30kN/m; deflection governs the quoted racking capacity. Deflection calculations follow from the formula presented in* **section 5.2**.

Calculate Wall Panel Deflection

From the equation presented previously:

$$\text{Deflection, } d_{max} = \frac{5.w.L^4}{384.E_t.I_p}.(1 + k_{def,t}) + \frac{w.L^2}{8.A_f.G_f}.(1 + k_{def,f}) \text{ ...where...}$$

Moment of Inertia of lintel, $I_L \approx 383x10^6mm^4$ *ignoring the OSB contribution*

Deflection, $d_{max(dead)}$ permanent condition
$= 5x1.6x1200^4x(1+0.8)/384/11{,}000/383x10^6 + 1.6x1200^2x(1+2.3)/8/11/2/400/1700$
$= 0.02 + 0.06 = 0.08mm$ ...*very low (noting also that shear deflection >> bending deflection)*

This lintel assembly, as loaded above, is adequate for a 1.2m span.

Chapter 5

Quick Note Summary

SIPs Wall Panels as a Deep Beam

Over one storey height (circa 2.5m) SIP panel walls exhibit deep beam behaviour, which may be utilised for carrying relatively light loads from roofs, cladding and finishes across spans up to 4m. However, in most circumstances, the saving that may be made from reducing the floor lintel is limited. In practice, it is prudent to ignore the wall contribution unless specific design criteria indicate an advantage from combining wall beam capability.

Where there are openings in the wall, this would reduce the overall stiffness and entail an increase in plane shear or racking. This is clear from the progressive collapse model in **section 3.6**, where the shear flow concentration is evident at the supports.

Notwithstanding the limited efficacy of walls as deep beams, the calculations and model demonstrate that the installed deflections reduce somewhat the design deflections of the supporting lintels.

A calculation demonstrated for the example in **section 5.2.2** that a 140x240mm glulam lintel would deflect long term 11.8mm. When built into the wall, the finite element model indicates that the long term deflection would be under 1mm.

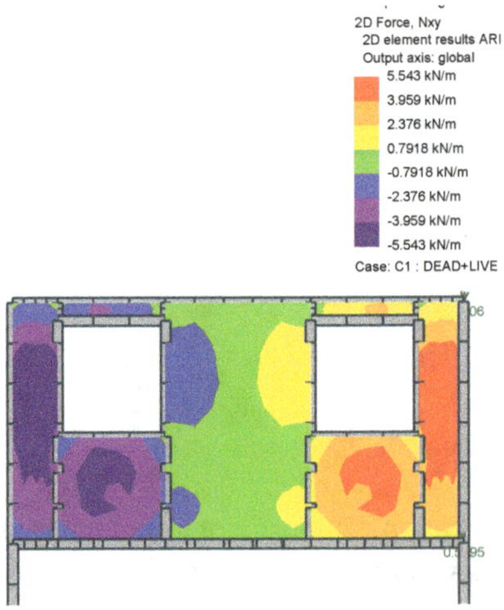

Figure 5.2.5 Racking from Section 3.6 Model

SIP Panel Lintels (OSB)

In **section 5.2.2** the 1.2m span lintel has shear flow or rolling shear, in the OSB faces, which is less than the design racking shear limits for an SBS panel, but may exceed the shear resistances of some types of panel. See **section 3.2.1** for permissible racking resistances.

For a 400mm deep section - which is typical for over large openings for a floor to floor height circa 2.6m and opening under 2.2m high - it highlights that this type of lintel will be adequate for short spans and for relatively light loading (such as when supporting roof loads alone).

SIP Panel Lintels (MgO)

Magnesium oxide (MgO) faced panels may be designed in a similar fashion as composite beams. Racking strengths of MgO panels are comparable with OSB faced panels. Manufacturer's data should be referred to for further information.

Chapter 6

SIPs as Cladding Support

SIPs as cladding Support. Cladding types. Test data for screw fixings. Batten support loading: dead and wind loads. Steel, concrete and oak frame details. Aspects of masonry support - ties, wall support. Brackets and rail details.

6.1 SIPS PANELS AS CLADDING INFILL AND CLADDING SUPPORT

Herein aspects of SIP as a structural wall or roof panel with clad external face, as shown in **figure 6.1.1**, are reviewed. Another application is also discussed where SIPs panels are not the primary structure, but which provide a structurally insulated clad external face fixed to the primary structure. In **figure 6.1.2**, SIPs panels are shown fixed to a vertical timber plate, bolted to a steel column. They are clad with a breather membrane and faced with ceramic cladding fixed to galvanised thin wall steel rails.

Aspects of the waterproof properties, insulation U value or fire resistance of the system is a consideration for architectural design. In light of the Grenfell fire disaster in 2017 in London, insurers are requiring engineers to specify if they are involved in cladding design. Design of SIPs panels by a structural engineer is undertaken with respect to SIPs as a structural component that is *part of the cladding*; their responsibility does not extend to the overall cladding system unless this is specifically called for by the client. Early stage clarification of this point is strongly recommended in case there is any doubt.

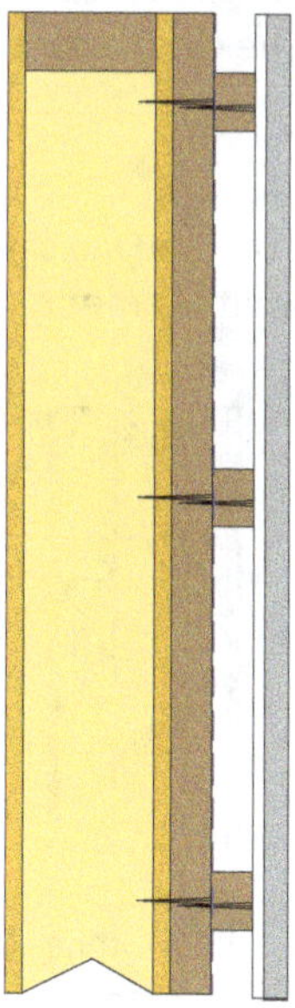

Figure 6.1.1 SIP Wall Clad with Render Board

Design for dead and wind loading on panels spanning horizontally or vertically is discussed.

Chapter Summary

Different cladding types, typical fixings and support is reviewed. Differential thermal movement or differential shrinkage or expansion is considered.

Fixings, with test data for screws into the facing, and some typical details are presented, along with different methods employed for attaching the cladding to the SIP face: treated timber battens, thin gauge steel brackets and masonry ties.

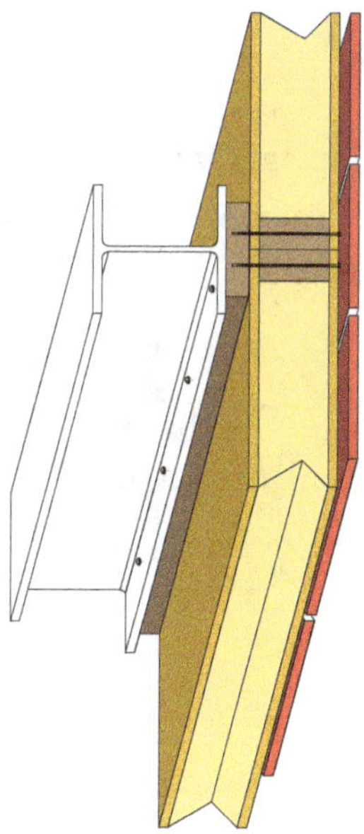

Figure 6.1.2 Steel Column with SIP Infill

6.2 SIPS PANELS AS CLADDING SUPPORT

Clad SIP wall and roof panels as shown in **figure 6.1.1,** must support the weight of the cladding when fixed with battens and also resist wind suction forces. Masonry, unless mounted as brick or stone slips on a board, is generally supported on an independent base or beam; tied to the SIP wall with screw fixed ties at regular centres. Standard tie spacing for masonry in SIPs and timber frame structures is 450mm vertically and 600mm horizontally, which equates to 3.7ties/m². See **section 6.3**.

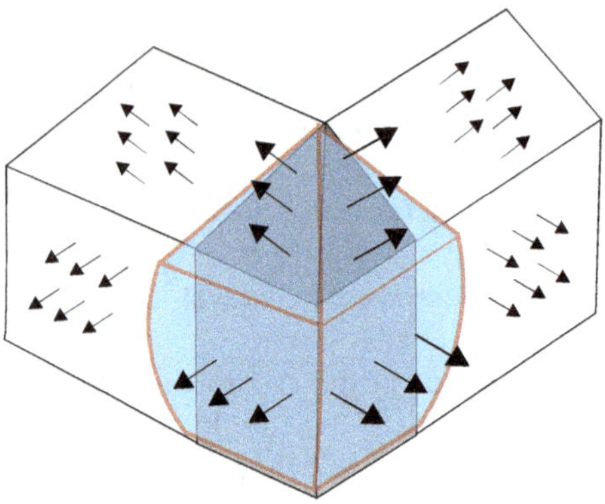

Figure 6.2.1 Wind Load Peak at Building Corners at Walls, Eaves and Roof Levels

When designing cladding support for walls, roofs and ties to resist wind loads, the engineer should consider the worst case local pressures, which are concentrated on the building corners, as illustrated in **figure 6.2.1**. Generally, the fixing configuration is checked for this worst case.

Vertical loads are carried by load sharing between lines of screws fixing the battens or rails to the SIP face. Battens are most commonly adopted in the common types of domestic cladding systems, with cold rolled thin wall brackets and rails for applications such as ceramic tiles or rainscreen cladding supplied as part of the cladding contract.

Typical Cladding Types for Walls and Roofs

In **figure 6.2.2,** typical suspended wall cladding types with the range of fitted indicative weights are shown, mounted on timber battens, fixed over the breather membrane.

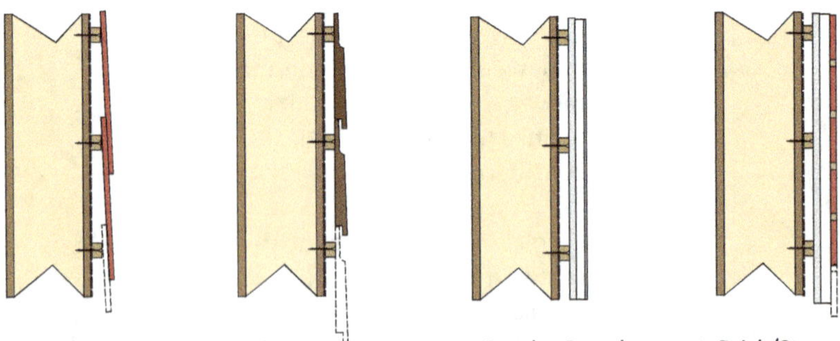

Figure 6.2.2 Typical SIP Wall Suspended Cladding Types with Indicative Weights

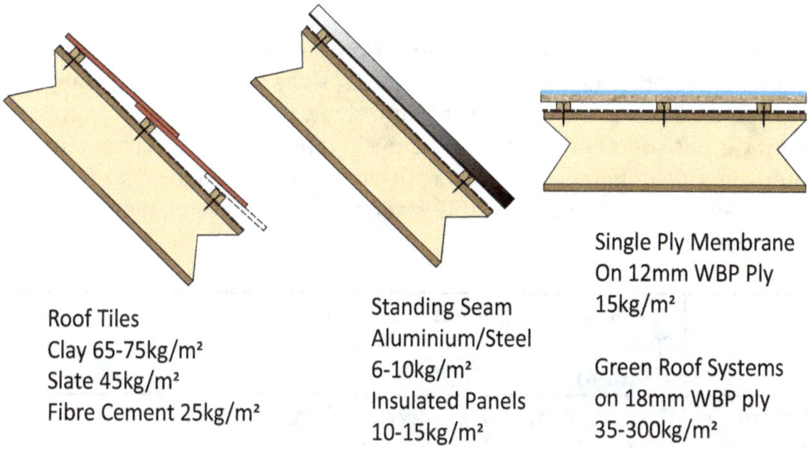

Figure 6.2.3 Typical SIP Roof Cladding Types with Indicative Weights

Roof Cladding

Roof cladding types with indicative weights are shown in **figure 6.2.3**. For flat roofs, mounting cladding on an external grade (WBP) ply deck on battens is recommended; combined with perimeter air outlets this achieves a vented space under. The ply on varied depth battens - or furrings - or SIP panels may be laid to a fall for drainage.

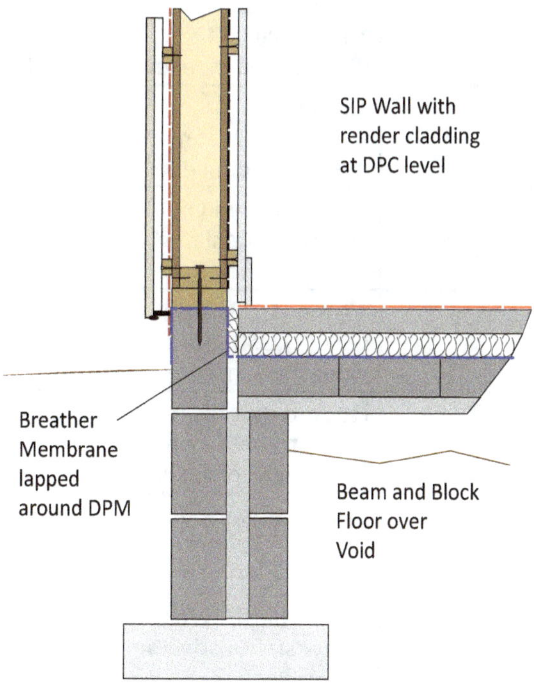

Figure 6.2.4 Wall Section At DPM Level

Green roofs vary considerably in weight depending on the system installed. Early confirmation of the client's preferred roof system is essential for the initial stages of design. SIPs with timber splines are appropriate.

Battens Supporting Cladding

Battens are laid in the direction suited to the cladding (vertically or horizontally in the case of a wall). If a large void is necessary between the SIP panel and cladding then occasionally wide battens up to circa 100mm may be employed. Treated timber is generally used since battens are in exposed conditions.

Cladding at Threshold Level

Figure 6.2.4 illustrates the integration of components of a clad SIP wall at DPM Level. Details are further presented in **chapter 8**.

6.3 CLADDING FIXINGS INTO OSB AND MgO FACES

Assessment of the screw fixings of battens or brackets into OSB or MgO panels cannot be calculated using Eurocode 5 methods, since there is a minimum for screw pointside penetration of 16 x diameter (3mm screw maximum for a 12mm face). Test data from various sources is presented below in **table 6.3.1**. A discussion of the best safe interpretation of figures for screw resistance in tension and shear follows, erring on the conservative. In most cases, this is adequate for typical suspended cladding systems.

Source	Pullout (Tension) kN (dia.)	Shear kN (dia.)	Comments
Structural Panels Australia	0.83 (4.2mm) 0.81 (6.3mm)	0.54 (4.2mm) 2.12 (6.3mm)	11mm OSB. Ultimate values
R-Control SIPS (USA)	0.82 (4.2mm) 0.79 (6.1mm)	0.55 (4.2mm) 2.11 (6.1mm)	11mm OSB (7/16") Average ultimate value (failure)
Tekhaus (now Kingspan) panel. JA3-6.5mm	1.0 (6.3mm)		15mm OSB. 5 samples. Mean - 3.7Sd (98% confidence level)
Kingspan JA3-6.5 EJOT	0.83 (6.3)	*Assumed to be same as tension value*	15mm OSB. 10 samples. Mean - 2.8Sd (98% confidence level)
Kingspan JA3-6.5 EJOT	0.90 (6.3)		15mm OSB 8 samples. Mean - 3.4Sd (98% confidence level)
CERAM 081118 5mm screws	0.63 (5mm)/ 0.77 (5mm)		11mm/15mm OSB Minimum declared values (*characteristic*). Bracket tests.
Headlok by Fastenmaster	0.85 (6.3mm)/ 1.87 (6.3mm)		12mm (16/32")/18mm (23/32) OSB . Minimum values assumed ultimate.
MgO Baumit Baseboard 10	0.89 *Screw diameter unspecified*		Quoted pull-through strength based on 81N/mm for 11mm thickness.
MgO Multipro Hanson BBA	0.38 (4.8mm)		Characteristic pullout through C16 timber from tests.
MgO Enviroboard fireboards	0.055 /0.087 *Screw diameter unspecified*	*Pro rata (top): 0.087x0.54/0.83 = 0.057*	9mm board/12mm Board screw size unspecified. *Shear not available: assumed value.*

Table 6.3.1 Pullout and Shear Text Data: Screws in OSB SIPs Panels

Derivation of Design Capacities for Screws in SIP Faces

Design shear and tension (pullout) values for 4.2mm screws, 5mm and 6.3mm screws, respectively #8 gauge, #10 and #14 gauge, are calculated below. It is assumed that critical failure is based on the facing material and not failure of the screws.

From **table 6.3.1,** the minimum *ultimate* values are assumed to be the equivalent of characteristic values in accordance with Eurocode 5:

4.2mm, 5mm and 6.3mm Screws Design Tension and Shear in 11mm OSB

Minimum tension resistance, $F_{ax,rk}$ = 0.63kN - *from 5mm Ceram 081118*
Minimum shear resistance, $F_{v,rk}$ = 0.54kN - *from 4.2mm Structural Panel Australia*

Tension resistance is related to wind suction, an instantaneous load with K_{mod} = 0.9 (service class 3) for OSB. Shear resistance is a permanent condition with K_{mod} = 0.3 (service class 3). For OSB, the material factor, Υ_M = 1.2.

Design tension resistance, $F_{ax,d} = F_{ax,d}.Kmod/\Upsilon_M = 0.63 \times 0.9/1.2 = 0.47$kN
Design shear resistance, $F_{v,d} = F_{v,rk}.Kmod/\Upsilon_M = 0.54 \times 0.3/1.2 = 0.14$kN

4.2mm and 6.3mm Screws Design Tension and Shear in 15mm OSB

Minimum tension resistance, $F_{ax,rk}$ = 0.74kN - *from 6.3mm EJOT tests by Kingspan*
Minimum shear resistance, $F_{v,rk}$ = 0.63kN (= 0.54x1.17*)
- **from 11mm OSB above: EC5, failure mode c, √15/√12 = 1.17*

Design tension resistance, $F_{ax,d} = F_{ax,d}.Kmod/\Upsilon_M = 0.74 \times 0.9/1.2 = 0.56$kN
Design shear resistance, $F_{v,d} = F_{v,d}.Kmod/\Upsilon_M = 0.63 \times 0.3/1.2 = 0.16$kN

4.2mm and 6.3mm Screws Design Tension and Shear in 12mm MgO

Pull through failure is ignored, as this is a failure of a screw/nail head fitted to the face of the board. The material factor Υ_M = 1.3, assuming consistency with particleboard.

Minimum tension resistance, $F_{ax,rk}$ = 0.087kN - *from Enviroboard panels*
Minimum shear resistance, $F_{v,rk}$ = 0.057kN - See **table 6.3.1**.

Design tension resistance, $F_{ax,d} = F_{ax,d}.Kmod/\Upsilon_M = 0.087 \times 0.9/1.3 = 0.06$kN
Design shear resistance, $F_{v,d} = F_{v,d}.Kmod/\Upsilon_M = 0.057 \times 0.3/1.3 = 0.013$kN

Comments on Design Screw Resistances

Values for 11mm and 15mm OSB are, in practical terms, the same, although it might be expected that a thicker face thickness is more robust in terms of the longer thread engagement. MgO board values are approximately 10% of the equivalent OSB boards, noting that cellulose fibre reinforced ResCom MgO boards have pullout values similar to OSB.

In terms of Eurocode 5 factors for multiple screws (n_{eff}) this should be taken into account for brackets where screws are in *close proximity*, in addition to observing minimum spacing. For batten fixings and masonry ties n_{eff} is assumed to be 1.0.

Example of Batten Fixing Checks in Tension and Shear

A wall with brick slips with a weight of 50kg/m² is fixed to horizontal 25x35mm deep battens at 400mm centres with fixings 400mm apart, as shown in **figure 6.3.1**. A design wind load suction of 1.5kN/m² occurs at the building corners. The resistance of the combined 4.2mm screw fixings into 11mm OSB is calculated below. It is convenient for clients to refer to cladding loads as weights, i.e. as kg/m².

Permanent Load Bearing Resistance (kg/m²)

Design load = 1.35 x 50kg/m² = 67.5kg/m²
Design shear resistance, $F_{v,d}$ = 0.14kN (14kg) per screw
Permanent resistance of wall fixings, $F_{p,v,d}$ = 14 /(0.4 x 0.4) = *87.5kg/m² > 67.5
**Equivalent to 0.88kN/m².*

Instantaneous Pullout Resistance (kN/m²)

Design wind suction, $P_{d,w}$ = 1.5 x 1.5 = 2.25kN/m²
Design tension resistance, $F_{ax,d}$ = 0.47kN per screw
Design suction resistance of wall fixings, $F_{t,d}$ = 0.47 / (0.4x0.4) = 2.9kN/m² > 2.25

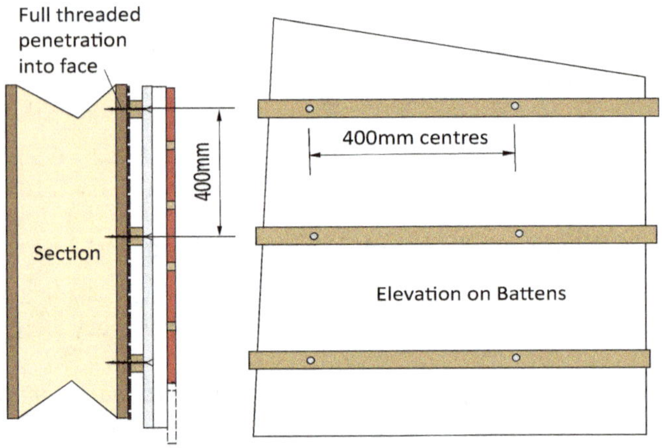

Figure 6.3.1 Wall Batten Check Example

MgO Board Fixings

As noted in the previous page, equivalent fixings into the MgO board have a fraction of the strength of OSB as a substrate. Battens or cold rolled steel rails need to be fixed to backing timber or cold rolled steel. Regular framing, such as timber studs or steel channels, are more appropriate for this application. Furthermore, screws should be *corrosion resistant* owing to the chemical interaction of the board in conjunction with water vapour permeation. Stainless steel or coating with a water proofing sealer is an appropriate specification. The engineer, with respect to MgO SIPs, should refer to the suppliers ' or manufacturers ' standard systems for fixings appropriate for the product.

Masonry or Stone Ties fixed to OSB SIPs Panels

For cavity wall construction with outer, independently supported masonry skin, ties may be fixed with screws to the outer face of the SIP panel. As noted in **section 3.2.4**, ties are commonly fixed with vertical spacing of 450mm and maximum horizontal spacing of 600mm in the horizontal bed joints, thus giving a tie distribution of 3.7 ties / m².

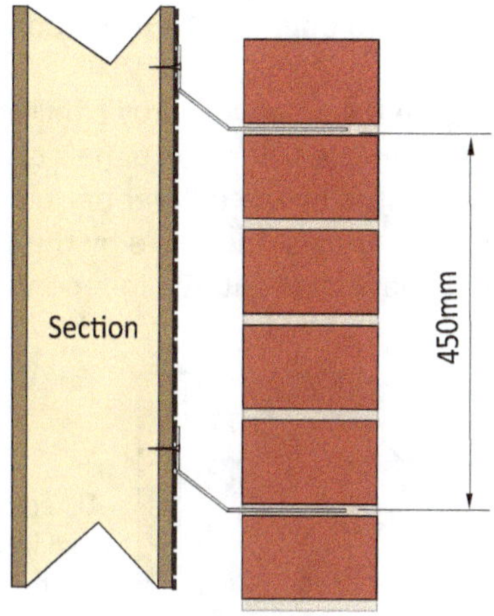

Figure 6.3.2 Masonry Ties

Assuming an instantaneous tension resistance of 0.56 kN from **section 6.3** for the screws in 15mm OSB, the design wind resistance is:

$F_{p,v,d(600)} = 0.56 \times 3.7 = 2.1kN/m^2$

For a locality with a high wind speed of circa 24m/s, for a building no greater than three stories, an external corner suction would be circa 1.2kN/m². The calculation below is on the limit of acceptability:

Design wind pressure,

$P_{p,v,d(600)} = 1.2 \times 1.5 = 1.8kN/m^2 < 2.1$

Masonry Ties Contribution to Racking in OSB from Section 3.2.5

To verify that a 100mm brick or block outer skin can contribute 25% of its weight to the racking dead load for a density of 3.7 ties / m² for 4.2mm screws, a calculation follows for instantaneous loading under wind loads. 11mm OSB is assumed.

Design shear resistance, $F_{v,d} = F_{v,rk}.K_{mod}/\Upsilon_M = 0.54 \times 0.9/1.2 = 0.41kN$
Shear contribution of 3.7 ties / m² = $3.7 \times 0.41 = 1.5kN/m^2$
25% of self weight of brick wall at 2.2kN/m² = $0.25 \times 2.2 = 0.55kN/m^2 < 1.5$

25% of masonry self weight is a reasonable assumption if the masonry is no more than 100mmm thick with up to 50% openings as a proportion of the wall area. For thicker walls it would be safe to assume the weight of the wall based on an assumed 100mm thickness.

Movement Joints in Brickwork

It is not usually the consideration of the SIPs designer to design the cladding. As a rule, extracted from LABC guidelines, brickwork should have vertical movement joints at every 10 to 12m but no more than 6m from a corner.

Leading onto the next section, which reviews SIPs as cladding to framed buildings, SIPs panels are generally supported off the frame, at each floor in new build construction, but if retrofitted to an existing frame can be self supporting to three stories as a rule. If the SIPs are brick clad, the differential vertical movement of brick, which expands over time, needs to be accommodated at every three stories height. In **figure 6.4.1,** a five storey concrete frame is retrofitted with brick clad SIPs. An intermediate frame is fixed to the side at the third storey to form a vertical joint.

6.4 SIPS AS A CLADDING SUPPORT IN FRAMED BUILDINGS

A useful application of SIP panels is when they are installed as an insulated support to cladding on framed buildings. They are also well suited to retrofitting framed buildings to upgrade the insulation and cladding.

In **figure 6.4.1,** a five storey concrete framed building was considered as a concept with retrofitted SIPs wall panels clad with an outer brick skin. The design was complicated by the differential expansion and contraction of the concrete frame and SIP walls and, furthermore by the differential expansion of the brick skin and SIP. This would have necessitated independent support on the third floor to the brick skin to support the skin on the upper two floors. A revised scheme with brick slips supported off the SIPs at each floor avoided the 3rd floor support and sliding brick ties. See **figure 6.4.2**.

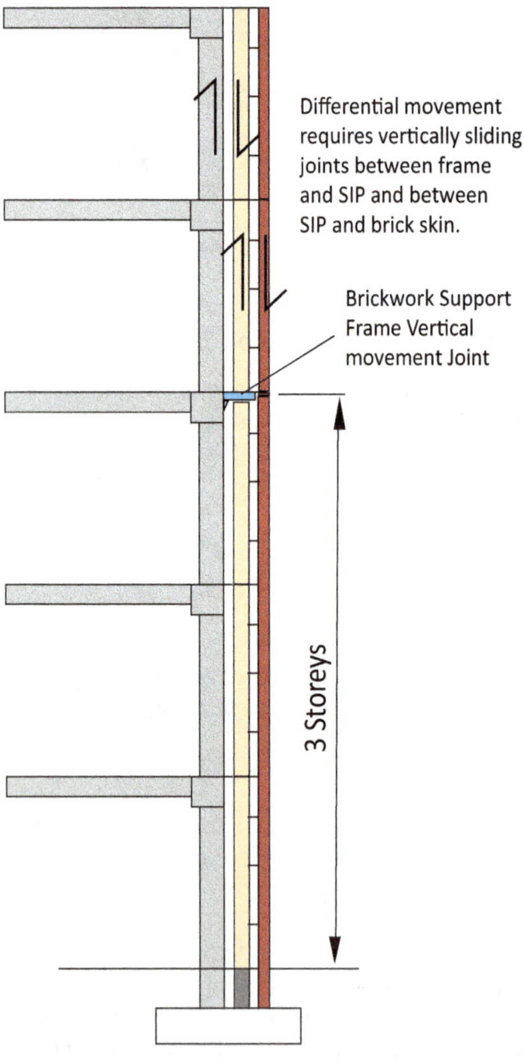

Figure 6.4.1 Concrete Frame Retrofitted with SIPs and Brick Cladding

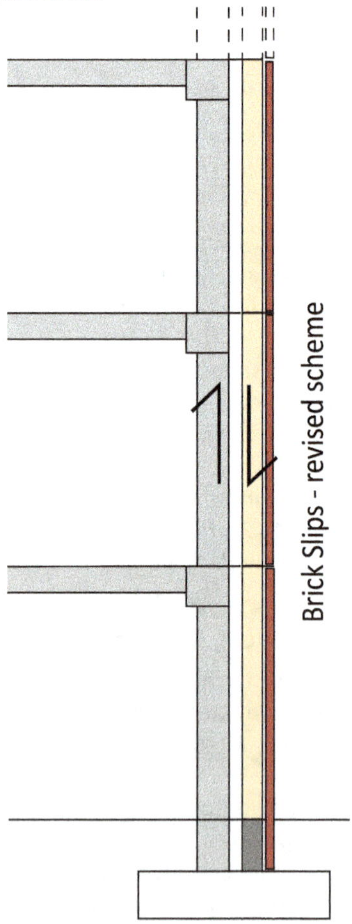

Figure 6.4.2 Concrete Frame Retrofitted Revised Scheme

SIP Panel Fixing to Concrete Frame

The SIP wall panels span vertically between the floors, as illustrated in **figure 6.4.3**, with vertically sliding bolted brackets that resist wind loading. Bolts have oversize washers to spread the local compression under wind suction. The brackets are fixed to each panel (1.2m spacing), designed to give adequate restraint.

The brackets are installed with resin anchors into the concrete frame with vertically slotted holes. The slots are extended, and the location of the brackets have tolerance to allow adjustment where reinforcing bars in the beam would impede the anchors. The SIP panels were checked for lateral wind loading and vertical loading accumulated over five stories. The galvanised steel brackets were checked for bending resistance and deformation under wind suction.

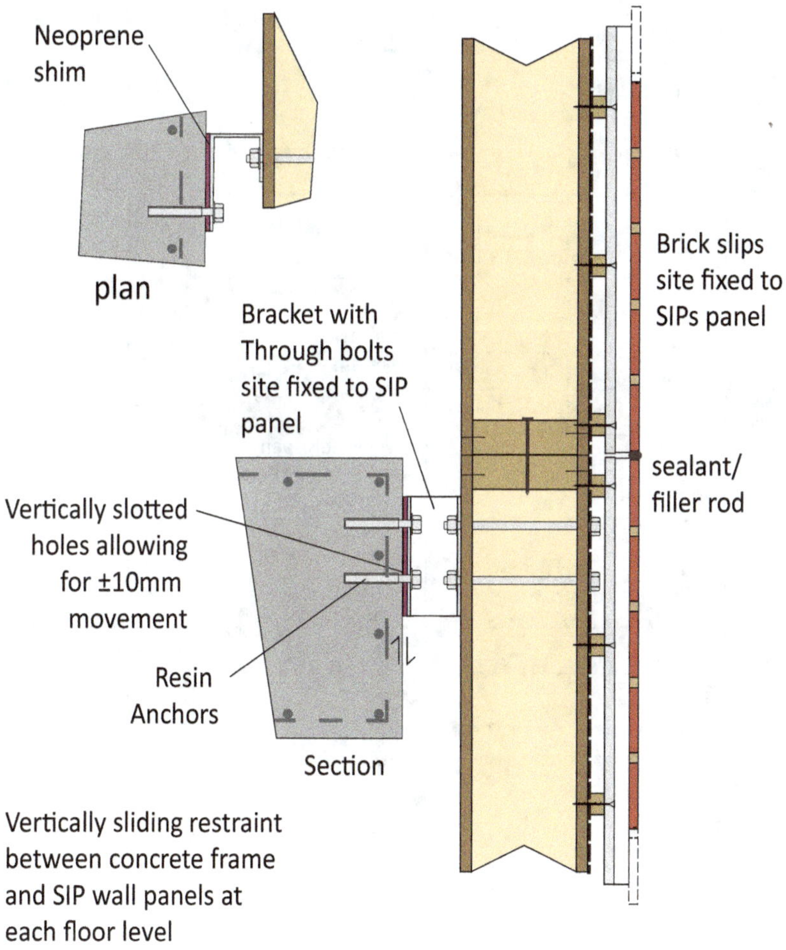

Figure 6.4.3 Vertically Sliding Brackets to Fix Each SIP Panel to Floor Beams

Example Calculation Concrete Retrofitted Frame (from Figure 6.3.2)

Calculations to check the 150mm retrofitted panels, elevated **figure 6.4.4,** are presented, based on the analysis methods outlined in **section 2.5**. With window openings present and a uniform wind suction of 0.8kN/m², the wind load is resisted by two 0.4m wide panels with 125x50 C24 solid timber splines on each side. A section through the panels resisting the wind load is shown in **figure 6.4.5**. The two panel sections resisting the wind load are checked as reinforced elements.

Design bending resistance in tension, $M_{d,t}$, combining OSB and timber rib, is based on the formula presented in **section 2.5**:

$$M_{d,t} = \frac{f_{t,o,k} \,.\, B_{eff} \,.\, t \,.\, K_{mod,f} \,.\, d_{eff}}{\Upsilon_{m,f}} + \frac{M_r \,.\, f_{m,k} \,.\, B_j \,.\, d_j^2 \,.\, k_{mod,j}}{6 \,.\, \Upsilon_{m,j}}$$

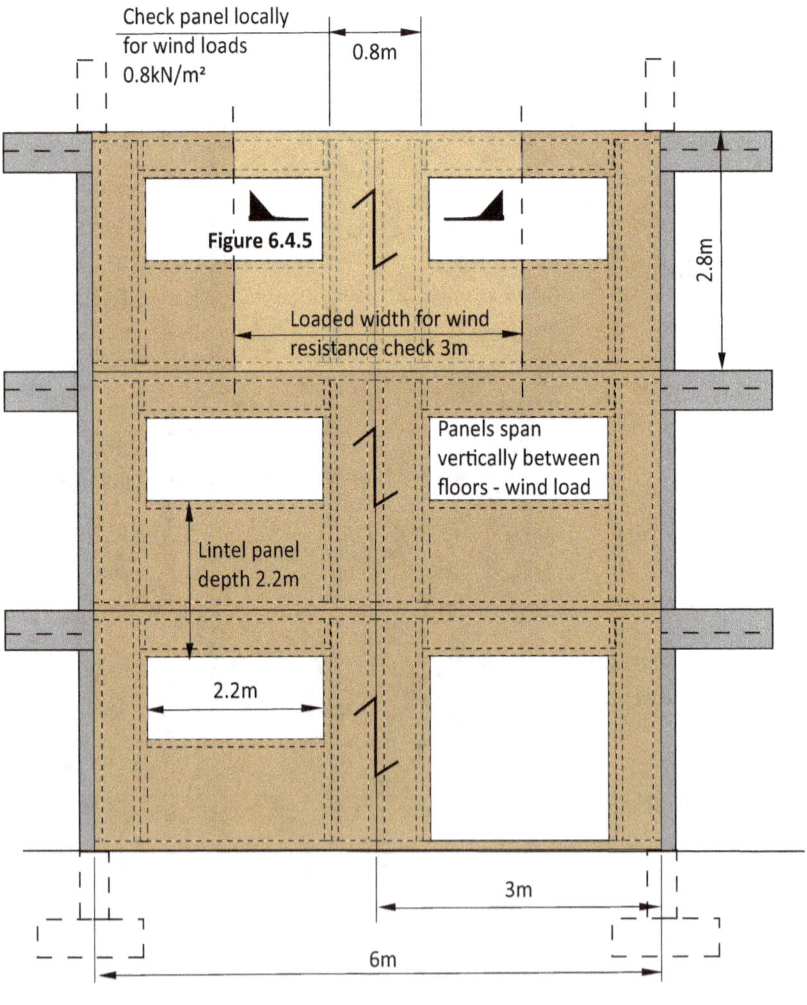

Figure 6.4.4 Part Elevation of SIP Retrofitted Panels Fixed to Concrete Frame

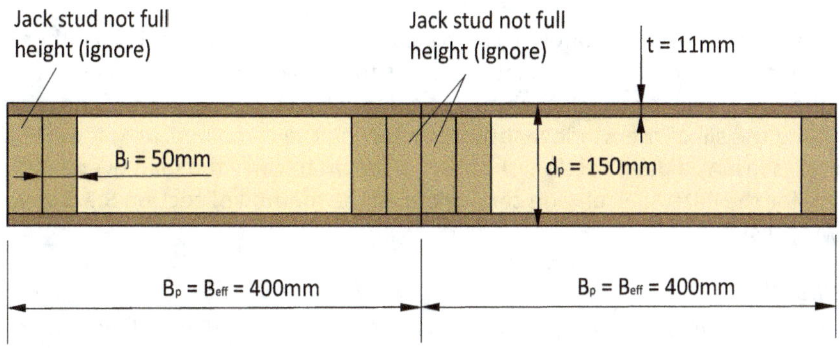

Figure 6.4.5 Local Section Through Central Wall Panels

*Check Bending Design Resistance (from **section 2.5**)*

Loading is instantaneous (k_{mod} = 0.9, OSB, k_{mod} = 1.1, timber). For a 400mm wide panel:

$M_{d,t}$ = 7.7 x 400 x 11 x 0.9 x (150-11)/1.2 + 0.35 x 24 x (2x50) x 125^2 x 1.1/6/1.3
$M_{d,t}$ = 3.53x10^6 + 1.86x10^6 = 5.44 x 10^6 Nmm = 5.39kNm

Where...$f_{t,0,k(OSB)}$ = 7.7N/mm², M_r = 3800/11,000 = 0.35 (C24 timber)

Applied design moment, M_d = 1.5w.h²/8 = 1.5x 0.80 x 1.5[ŀ] x 2.8² / 8 = 1.76kNm < 5.39
[ŀ]*Note that the loaded width for a 400mm wide panel is 3m/2 = 1.5m*

*Check Instantaneous Deflection (from **section 2.6**) K_{def} = 0*

$$d_{max} = \frac{5.w.h^4}{384.\{E_f.I_f/(1 + k_{def,f}) + E_j.I_j/(1 + k_{def,j})\}} + \frac{w.h^2}{8.A_j.G_j}.(1 + k_{def,j})$$

$I_f = (d_p^3 - d_c^3).B_{eff}/12 = (150^3 - 128^3).400/12 = 42.6 \times 10^6$ mm⁴
$I_j = B_j.d_j^3/12 = (2 \times 50) \times 125^3 / 12 = 16.3 \times 10^6$ mm⁴
$A_j = B_j.d_j = (2 \times 50) \times 125 = 12.5 \times 10^3$ mm²

$d_{max(B)}$ = 5x0.8x2800⁴/384/{3800x 42.6 x 10⁶/1 + 11,000x 16.3 x 10⁶/1} = 1.9mm
$d_{max(v)}$ = 0.8 x 2800² / 8 / 12.5 x 10³ / 690 = 0.1mm

dmax = $d_{max(B)}$ + $d_{max(v)}$ = 1.9 + 0.1 = 2.0mm (h/1400) - *deflection is low*
The central panels combined are adequate to resist the wind loading.

Connection Bracket to Concrete Beam (at 1.2m spacing)

Design wind reaction at floor level, $R_{brk,d}$ = 1.5 x 0.8 x 2.8 x 1.2 = 4kN (2kN per bolt)
From **figure 2.12.6,** pull through a resistance of a screw head is 4.7kN > 2kN.

*Note: a M10 bolt has a diameter of 18mm, which, with a washer, exceeding the presumed 15mm diameter of the screw example in **figure 2.12.6**, i.e., conservative.*

*Lintels as SIP Panels (from **section 5.2.2**)*

Lintel panels span approximately 2.2m, based on 400mm wide panels on either side. The client requested no intermediate brackets to reduce the cold bridging. In **figure 6.4.4** the lintels may be checked assuming a depth between the head of the window below and the sill of the window above, supporting the combined weight of SIP panel and brick slips. As a deep section, this can be shown to carry the load over a depth of 2.2m for the lintel. Calculation carried out to the method of **section 5.2.2** gave the following results:

Calculate Design Bending Moment and Support Shear

In **figure 6.4.3** the loads acting on the wall are based on the SIP panel (w_p) and brick slip self weight (w_{bs}) at 50kg/m². Header and bottom plate are 125x50mm deep (C24).

Total dead load, $W_{dead} = W_p + W_{bs} = 2.2x\ (0.25 + 0.50) = 1.65$kN/m
Total live load, $W_{live} = 0$

Design load, $W_d = 1.65 \times 1.35 = 2.2$kN/m
Design Moment, $M_d = W.L^2/8 = 2.2 \times 2.2^2 / 8 = 1.33$kNm
Design Shear, $V_d = W_d.L /2 = 2.2 \times 2.2 / 2 = 2.4$kN

Calculate Compression in Header and Tension in Bottom Plate

Axial forces acting on the header and first floor lintel are calculated from their centroid position from the neutral axis, which is central. See **section 5.2.2**.

Compression in 50mm header, $F_h = M_d / \bar{Y}_{top}$ ... $\bar{Y}_{top} = (2200\text{-}50)/2 = 1075$mm

Tension in bottom plate, $F_{bp} = 1.33/1.075 = 1.2$kN (= F_h *compression*)
Tensile stress in bottom plate, $f_{tbp} = F_{bp}/A_{bp} = 1.2x10^3 /125 / 50 = 0.19\text{N/mm}^2 << 8.6$
...where $f_{t,0,d} = 14x0.8/1.3 = 8.6\text{N/mm}^2$

Calculate Maximum Shear Flow (Rolling Shear) in Lintel

Shear flow at position of header, $V_{ph} = \dfrac{V_d.\ A_h.\ \bar{Y}top}{2.I_p}$ ...where...

Moment of inertia of panel beam, I_p:
$I_p = A_{sp}.\ \bar{Y}_{bottom}^2 + A_h.\ \bar{Y}_{top}^2 = 125 \times 50 \times 1075^2 \times 2 = 14.4x10^9\text{mm}^4$

Shear flow at position of header, $V_{ph} = 2.4x10^3 x\ 125x\ 50 \times 1075/2/14.4x10^9 = 0.6$kN/m
This is a low shear flow (rolling shear).

Calculate Wall Panel Vertical Deflection

From the equation presented in **section 5.2.2**:

$$\text{Deflection, } d_{max} = \frac{5.w.L^4}{384.E_t.I_p} . (1 + k_{def,t}) + \frac{w.L^2}{8.A_f.G_f} . (1 + k_{def,f}) \text{ ...where...}$$

$d_{max} = 5x1.65x2200^4x1.6/384/11{,}000/14.4x10^9 + 1.65x2200^2x3.3/8/11/2200/2/1700$
$= 0.01 + 0.04 = 0.05$mm ... *very low*

Horizontally Laid Panels

In **figure 6.4.6** wall panels are shown laid between supporting posts. Sometimes, it is advantageous to take advantage of the SIP panel length for wind resistance, such as when the post height is greater than 4m, and the post spacing necessitates a large eaves beam to resist wind loading if the panels were vertically installed.

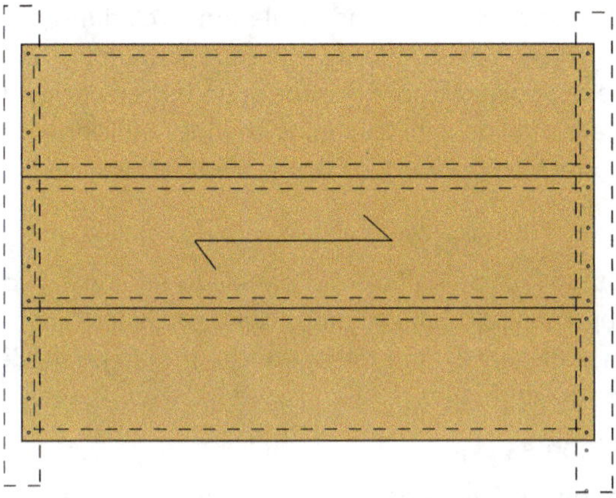

Figure 6.4.6 Horizontally Laid Wall Panels

At 4m width, the panels, under instantaneous loading, may be shown to be adequate without timber reinforcing splines.

Vertical loading may be minimised if the frame is designed to support the SIP wall and roof panels. Verification would follow the same principles as the previous example for panel wind resistance.

SIP Clad Oak Frames

Oak framed houses are another type of timber framing which are combined with SIP external walls and roofs, enabling a well insulated contemporary house with a historic vernacular expression.

In **figure 6.4.7**, typical wall fixings are illustrated. SIP roof and wall panels were attached to oak posts and beams. The SIP supplier and erector fitted the panels to the built oak frame. Where timber battens and plates were fixed to oak stainless steel screws were specified to avoid degradation from the oak tannin.

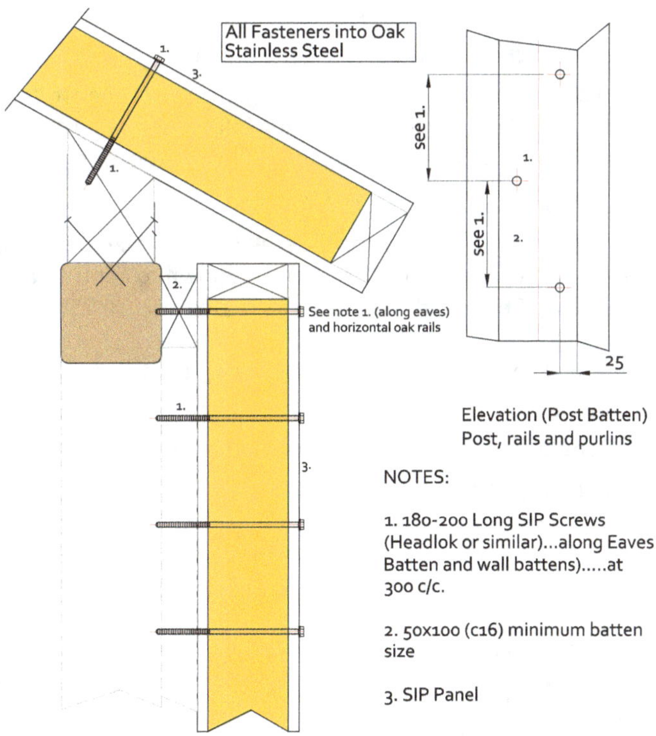

Figure 6.4.7 Oak Frame Wall Panel Fixings

Chapter 6

Quick Note Summary

SIP Walls as Infill and Cladding Support

SIP panels are ideally suited to provide robust support and insulation to cladding, whether as an independent structure or as an external skin to a framed building. A wide variety of cladding types may be supported from SIP panels, on battens or cold rolled steel rails or brackets. Cavity walls may be constructed with inner SIP skins and outer skins in masonry, brick, rendered blockwork or stone, tied back to the SIP panels.

Commonly, battens are fixed to the outer face of wall or roof panels by screws, which can support the weight of the cladding and wind load suction. Wind pressures are greatest at the corners of buildings in walls, eaves and roofs. Checks should be made in these locations for any cladding design checks.

Screw Fixings to the Face of SIP Panels

Although it is not possible to verify fixings to Eurocode 5 owing to the pointside penetration into a typical wall face, test data is shown in **table 6.3.1** sufficient to derive design capacities for typical screw fixings. It is necessary to verify that the screws are capable of resisting instantaneous wind loads and the long term dead load from the supported cladding.

MgO SIPs Screw Fixings

Magnesium oxide board faces have significantly lower screw capacities than OSB faced panels. Supplementary framing in timber or thin gauge steel can provide cladding support. Screws should be stainless steel or protective coated.

Masonry Ties and Movement Joints

Masonry ties to SIPs faces are generally at closer centres than masonry construction, typically 3.7 - 4.4 ties/m^2. They may be shown to contribute to racking resistance as a proportion of the dead weight of the masonry, circa 25% of the weight of the skin.

Differential vertical movement needs consideration for buildings greater than three storeys high.

SIP Panels as External Walls to Framed Buildings

SIP panels are frequently installed as the external wall or roof of framed buildings; whether new construction or retrofitted to existing frames. Differential horizontal and vertical movement is a consideration. Generally, panel assemblies are split so that they are compartmentalised within individual floors and bays.

Panels may span vertically or horizontally since the dead and live loading is limited by the weight of the panels and supported cladding. Lintels may be designed as SIP beams.

Oak framed buildings are often clad with SIP panels, taking advantage of their insulative properties.

Chapter 7
Modular Construction

Types of modular construction: site assembled, partial pre-assembly and built wholly offsite. Small units: independent, extensions and temporary. Floor chassis. Multistorey: houses, flats and offices. Lifting, craning and assembly.

7.1 TYPES OF MODULAR CONSTRUCTION

This is a form of construction that is based on the assembly of standard units in an offsite factory, which is then transported whole or as partially built components that are assembled on site. When transported as whole units, there are dimensional limits to width, height and length; in the UK up to 3.5m in width and 18m in length as a guide. When transporting a unit or load over 2.9m wide and longer than 18.65m, which is classed as an abnormal load, special permits are needed. An important consideration is whether narrow roads access the site or the site is problematic for lifting by crane or accessing the specific location. For this reason, modular units are frequently transported as components for full or partial site erection. Units may be single storey or multi storey, installed on site within supporting framed structures.

In **section 1.3.2** SIPs modular accommodation units for a charity project are shown being assembled in a workshop in **figure 1.3.5**. The units have steel framed chassis and joisted roofs with SIP external walls. The unit in **figure 1.3.6** was site installed, built as fully SIPs roof and floor with timber trimming. The roof 'V' shaped components were fitted onto 'U' shaped floor and wall assemblies and located in place with OSB strips; connected in such a way as to provide structural stability.

The unit shown in **figure 7.1.1** was a concept design in which 'portalised' timber frames with steel plated, moment resisting, splice joints would be infilled with SIP panels, supported off screw piles.

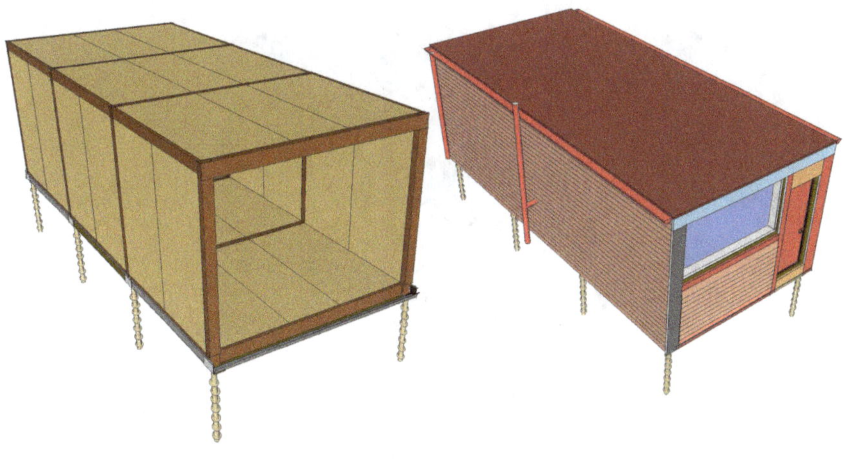

Figure 7.1.1 Concept Modular Unit SIPs Integrated with Timber Frames

7.2 COMBINING MODULAR UNITS

Side by Side Units

A variation on the concept shown in **figure 7.1.1** is shown below in **figure 7.2.1**. Monopitched units with open sides are connected together on site with central posts as part of the portal framing to form a double width unit of 'bungalow' size. Units have been built for this type of construction with back to back modules, which are stabilised by racking walls internally and externally without portal frames.

Figure 7.2.1 Concept Modular Unit Back to Back Monopitch Frames

Multistorey Units

Multistorey units, which are typically built within supporting structural frames, take advantage of the speed of construction and offsite manufacture of modular units. **Figure 7.2.2** shows a multistorey concept based on the units in **figure 7.1.1**.

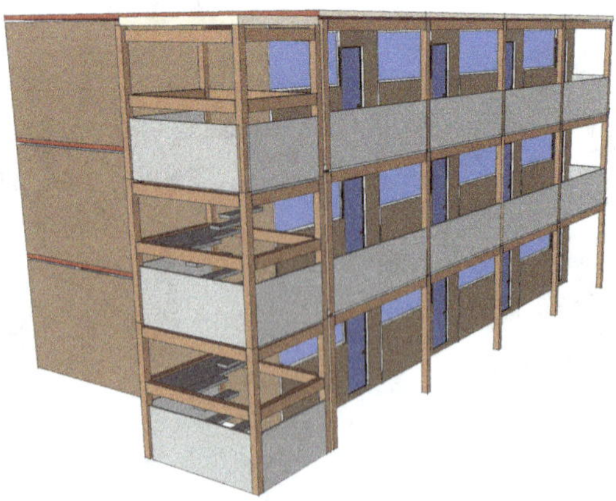

Figure 7.2.2 Concept Modular Units Multistorey

7.3 TYPICAL GARDEN ROOM/SUMMER HOUSE MODULAR UNIT

A typical garden room is shown in **figure 7.3.1**. For clarity, the cladding is not shown. This unit is 3.3m wide, unclad, by 4.5m long and 2.5m in height clad. See **figure 7.3.6**. According to UK building regulations, planning permission is not needed if the unit is no more than 2.5m high and no more than 30m² in area, unless in a conservation area. The floor is a cold rolled steel cassette as an alternative to SIP or timber. This is more resistance to the risk of dampness, which would be a matter of concern for a SIP floor. See **figure 7.3.2,** showing the cold rolled joists and timber splines in the SIP roof panels.

Figure 7.3.1 Typical Garden Room Modular Unit

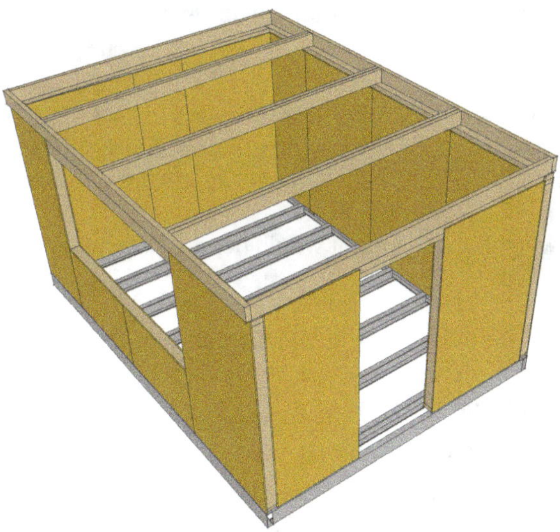

Figure 7.3.2 Garden Unit. Floor Deck and SIP Roof Hidden

The floor deck and roof panels are hidden for clarity.

The unit is no more than circa 2 tonnes unclad, and 6 tonnes with cladding which equates to 20kN and 60kN, respectively. Assuming the unit sits on four corner plinths on the long edge, this gives a foundation dead load of circa 15kN under each support, which is achievable with a concrete slab or simple footings.

7.3.1 INSTALLATION/ERECTION

In **figure 7.3.3,** a garden room unit is shown being craned into position with bolted face plates which are fixed to lifting cables. Two lifting beams above enable balanced hoisting of the unit.

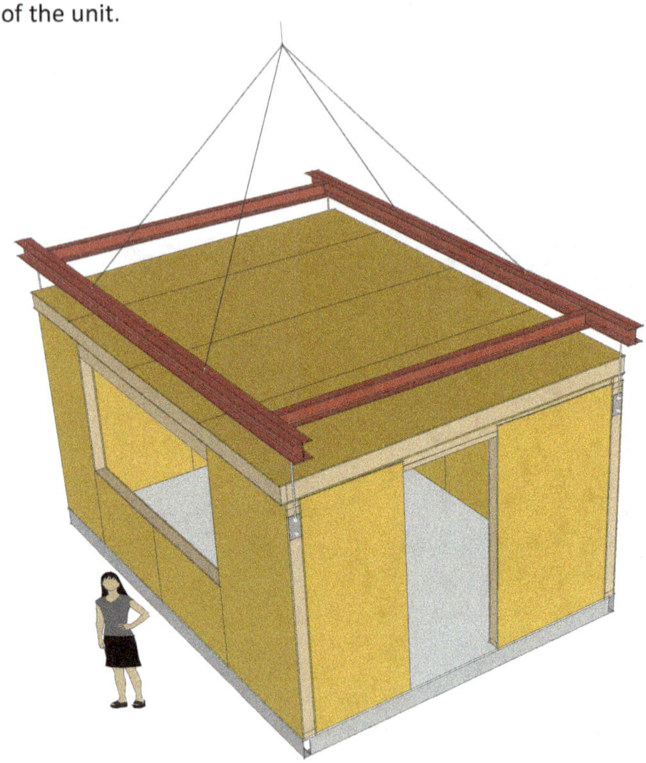

Figure 7.3.3 Crane Lifting Arrangement for Single Unit

Loading Checks on Unit

As noted previously the unclad unit weighs 2 tonnes, which exerts a load at each lifting point equivalent to 5kN. Allowing for a dynamic load factor during lifting of 1.5, that requires each bolted plate to carry 7.5kN as a design load (3.75kN per bolt).

Note that if the unit was clad, weighing 6 tonnes, it would need a stronger fixing arrangement that reviewed overleaf.

The corner posts are 100mm wide at the corners, to enable a greater lifting strength for a bolted joint. M16 bolts in pairs connect the 6mm plates through the corner posts. A summary of the loading and pertinent dimensions are shown in **figure 7.3.4**.

Calculate Resistance of Bolted Plate Connection

In **figure 7.3.5,** the bolted plate connection is shown with applied design loads for each bolt. It was found from the Eurocode method for serviceability class 3, with *instantaneous* loading, that the design shear resistance per bolt, $f_{ax,0,d}$ with a 6mm steel plate is 10.7kN for loading parallel to the grain.

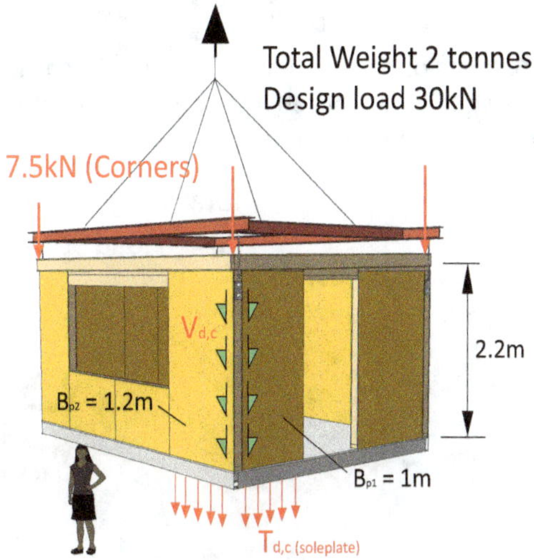

Figure 7.3.4 Lifting Loads on Unit

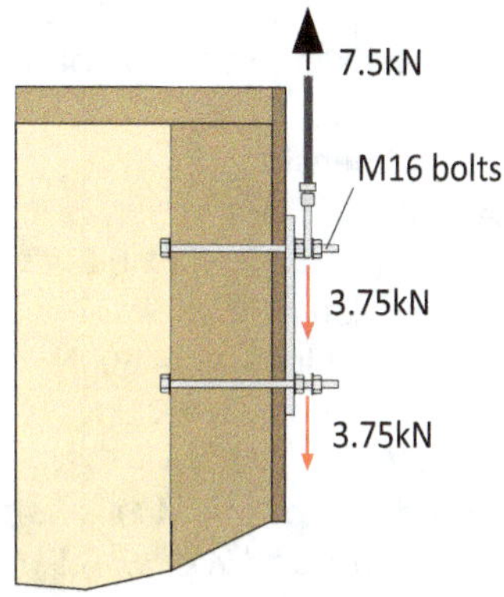

Figure 7.3.5 Bolt Shear Forces

Check Bolt Pair Resistance

From Eurocode 5, section 8.5, the effective number of bolts, η_{eff}, in a multiple bolt arrangement is calculated from the *minimum* of the two values below:

$\eta_{eff} = n = 2$
$\eta_{eff} = n^{0.9}. \sqrt[4]{(a_1/13d)}$ ...where a_1 = bolt spacing *parallel to grain* and d = bolt diameter
$\eta_{eff} = 2^{0.9}. \sqrt[4]{(100/13/16)} = 1.87 \times 0.83 = 1.55 < 2$

Therefore, η_{eff} (min) = 1.55

Resistance of the bolt pair, $f_{ax,0,d(pair)}$ (for $f_{ax,0,d}$ = 10.7kN), assuming a k_{mod} = 1.10 *instantaneous* and material factor, Υ_M = 1.3:

$f_{ax,0,d(pair)} = \eta_{eff}. f_{ax,0,d}. k_{mod} / \Upsilon_M = 1.55 \times 10.7kN \times 1.1/ 1.3 = 14.0kN > 7.5kN$

Calculate Panel In-Plane Shear (Racking)

For the corner panel at 2.2m height, sharing the corner load:

Panel racking shear, R_{rk} = 7.5kN/2.2/2 = 1.7kN/m < 5.4 (SBS panel from **section 3.2.1**)

Calculate Uplift on Soleplate Connection to Cold Rolled Steel Chassis

From **figure 7.3.4,** line loading is assumed, in the worst case, to be spread over the corner with adjacent openings.

Design tension line load, $T_{d,c} = 7.5 / (B_{p1} + B_{p2}) = 7.5 / (1.0+1.2) = 3.4$kN/m
Assuming a typical characteristic tension resistance, ft,d, for self tapping screws of 2kN in thin gauge steel, and 150mm spacing, and Υ_M = 1.0 for steel:

$f_{ax,d(soleplate)} = 2kN. k_{mod}. (B_{p1} + B_{p2})/ 0.15 / \Upsilon_M = 2 \times 1.1 \times 2.2 /0.15/1.0 = 32.3kN >> 7.5$

7.3.2 STABILITY ANALYSIS OF A BUILT GARDEN OFFICE

A fully clad unit is shown in **figure 7.3.6**. It sits on four pad footings on each corner, 0.5m x 0.5m in plan, with a galvanised base plate anchored into the concrete at four points, fixed to the side of the floor cassette with 8 self tapping screws.

Plan dimension, (L x W_d)
= 4.5m x 3.3m
Door dimension = 2 x 0.9m wide
Height clad = 2.5m
Total dead load, G_k = 60kN
Total live load, Q_k = 31kN
Total Wind Load, W_k = 9kN
(Wind load at top, W_{kT} is half the total wind load W_k i.e., 4.5kN)

Vertical Reactions

Live load, $R_{v(Qk)}$ = 31/4 = 7.8kN
Dead load, $R_{v(Gk)}$ = 60/4 = 15kN

Wind load (windward/Ww) reaction
$R_{v(ww)}$ = -W_{kT}.h./2.W_d
= -(1x4.5)x2.5/2/3.3 = -1.7kN *uplift*

Wind load (leeward/lw) reaction
$R_{v(LW)}$ = -$R_{v(ww)}$ = 1.7kN

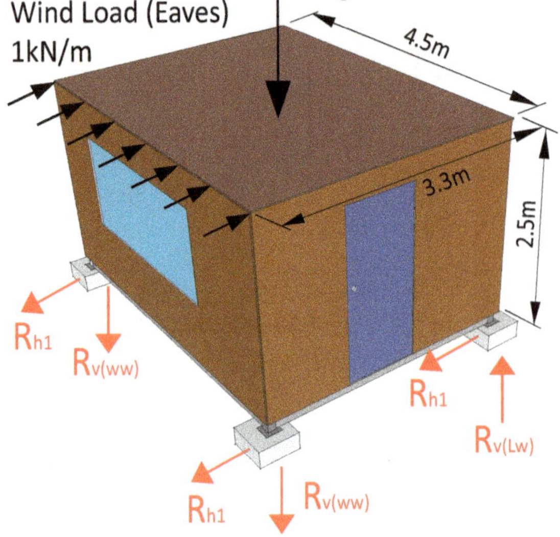

Figure 7.3.6 Clad Unit with Wind Load (Only)

Combined Vertical Reactions (at Each Base)

Dead load + Live Load = $R_{v(Qk)}$ + $R_{v(Gk)}$ = 60/4 + 31/4 = 15 + 7.8 = 22.8kN
Dead Load + wind (ww) = $R_{v(Gk)}$ - $R_{v(ww)}$ = 15 - 1.7 = 13.3kN *(i.e. no net uplift)*
Dead Load + wind (lw) = $R_{v(Gk)}$ + $R_{v(ww)}$ = 15 + 1.7 = 16.7kN (*weight of base ignored*)

Horizontal Reactions from Wind

R_{h1} = 9kN / 4 = 2.25kN at each base

Maximum Ground Bearing Pressure on Pads (0.6x0.6m) from Above:

Bearing pressure, P_b = ($R_{v(Qk)}$ + $R_{v(Gk)}$) / 0.6^2 = 22.8/.36 = 63kN/m^2
Adequate for reasonably firm ground, assuming confirmation by the engineer.

Check Overturning - from Section 3.2.6

F_{OT} at eaves = 4.5/2 = 2.25kN
M_{OT} = 2.25 x 2.5 x 1.5 = 8.4kNm (each gable)
M_R = 60/4 x 3.3 / 2 = 24.8kNm >> 8.4 (*ignoring the footing weight - conservative*)

Racking on Gable with Door (Worst Case) - from Section 3.2.7

K105 = $(2.5/2.4)^{0.4}$ = 1.02, K106 = $(1 - 1.3x\{2x.9/(2.5x3.3\})^2$ = 0.51,
K107 = 1.0 (*assumed, conservative*), K108 = 1.0 (*no internal walls*).

Wall racking resistance, R_v = 5.4kN/m. Unclad wall Length, L_{unc} = 3.1m

Racking resistance, R_{rk} = 5.4 x 3.1 x 1.02 x 0.51 x 1.0 x 1.0 = 8.7kN >> 2.25

7.3.3 TRANSPORTATION

As discussed in **section 7.1** transportation and installation of units is an important consideration. Transport costs and options for carriage are most economical if the units are transported with panels, timber and floors broken down into individual parts. This has the disadvantage that this requires full site assembly. Many clients are happy with this arrangement, allowing them to have control of site costs, which can be handled on site with no cranes required.

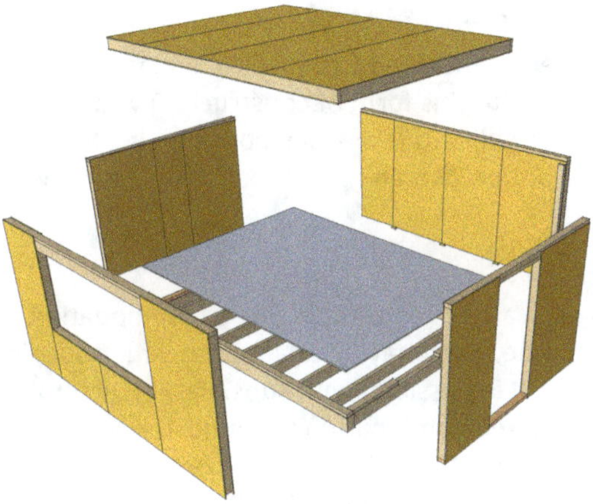

Figure 7.3.7 Exploded View of Panels with Timber Cassette Floor

Figure 7.3.7 shows an exploded view of the components, which are partly assembled so that they can be transported, as shown in **figure 7.3.8** (left), prior to restraint straps being fitted.

If the panels can be fitted, leaning on their sides in a container or sheet sided wagon this would avoid abnormal load restrictions.

For minimal site construction, whole units could be transported; options ranging from unclad to fully clad and fitted out with services, ducting and cabling; and even fully fitted out kitchen, bathroom and other units.

Figure 7.3.8 Units in Transit - Wall Components and Fully Assembled

Chapter 7

Quick Note Summary

Modular Construction

Modular construction can offer high quality factory built, partly assembled or fully assembled units, which can be transported to the site by road with minimal site erection and fitting out. SIPs is ideally suited to this form of construction with its high level of insulation and airtight fit. It offers rigidity in terms of stability owing to its racking strength, which also enables the option of craning fully assembled units with minimal temporary bracing.

Types of Modular Construction

Single unit modules are ideal for units such as garden rooms, site accommodation and holiday lets. Combined together, side by side, they may be adapted to larger buildings such as bungalows or single storey offices. It is possible, with subsidiary structural framing, to combine multiple units in a stacked combination for multistorey buildings. Student accommodation commonly adopts offsite fabrication for ease and speed of site erection. Open sided units are frequently adopted for building extensions to existing buildings.

Lifting Structural Considerations

Structural design should allow for strengthening, if necessary, to allow for lifting by crane of whole or partially built units. Some designs adopt steel corner posts, which can be fixed to floor trimmers or cassettes, allowing easy access for lifting equipment connection by shackles or other means. The clad weight may have to be considered, which can be typically threefold or more greater than the unclad weight.

Structural Design

Structural design of floor trimmers and joists may have to take into account large openings or fully open sides. Roof panels and wall panel fixings and design is similar to larger scale SIP buildings. Racking needs to be checked, with framing possibly needed at large openings for stability purposes.

Foundations are commonly simple concrete pads arranged to suit the spanning capacity of floors or screw piles with ground beam systems or concrete slabs with concrete plinths or masonry sleeper walls.

Stability calculations should include uplift and sliding checks from wind loading. For single storey clad units this is often not of concern. On occasion, in high wind, exposed locations, heavier foundations or additional strapping may be needed.

Transportation

Deciding on transport options must take into account the roads, crane availability and site access. Whether the supplier and client opt for fully built units offsite or, at the other end of the scale, broken down into the smallest components will depend on a trade off between site erection, time limitations and cost. Whole units, factory built, require greater transport costs but minimal site preparation and erection in terms of time and site labour.

Chapter 8
Connection Details

Various typical connection details. Soleplates at DPM level: anchors and straps. Upper external wall/floor junctions. Eaves, ridge and hip details. Screw and bolts fixings, typical capacities. Timber, steel and flitch beam at wall supports.

8.1 TYPICAL SIPS CONNECTION DETAILS

Manufacturers generally provide standard details for SIPs panels, which cover a range of typical connections at floor and roof junctions, splice details, soleplates and so forth. Manufacturers publish structural data in their BBA or ETA approved certification for SIPs systems. Test data, which is used to compile these documents, is based on specific fixings. For example, basic racking capacities are obtained from tests on panels with soleplates and headers fixed with nails and adhesives or silicone sealants unique to that manufacturer's SIP construction system.

Throughout the book, details pertinent to the structural methods discussed are shown in sufficient detail to guide the reader through the calculations. In this section details will be shown in more detail with options shown where applicable. It is not an exhaustive guide. The intention is to provide enough generic information to assist the engineer in progressing with their own details.

It would not be possible to cover all aspects of SIPs structures since each project will have unique design problems to resolve, many of which are driven by architects and clients. For instance, the avoidance of cold bridging where a steel floor trimmer sits within an external wall, or the demands of a 'Passivhaus' low energy accredited project.

8.2 HIGHLIGHTING STRUCTURAL AREAS OF CONCERN

Footnotes will be provided where detail is of interest with reference to the design subject areas in previous chapters. For example, a soleplate detail at DPM level shows various options for support, such as concrete plinths or slabs and block supporting walls, is particularly of interest in the course of checking the racking resistance of walls, which is outlined in **chapter 3**.

A first reference for the engineer is to obtain the specific product manufacturer's details manual. The suppliers and builders will work with the general specifications and fixings schedules unless advised otherwise. The details shown therein should, therefore, be taken as examples that are generic, and, if used, adapted to different products such as Hemsec or Kingspan panel systems.

With respect to **figure 8.3.1**, showing fixing spacing, the rules are simplified. The engineer should consult Eurocode 5, section 8, to check as to whether closer spacings are possible for the specific conditions; relevant references for edge distances and spacing: nails (Table 8.2), bolts (Table 8.4), dowels (Table 8.5) and screws (Table 8.6).

8.3 INDEX OF CONNECTION DETAILS

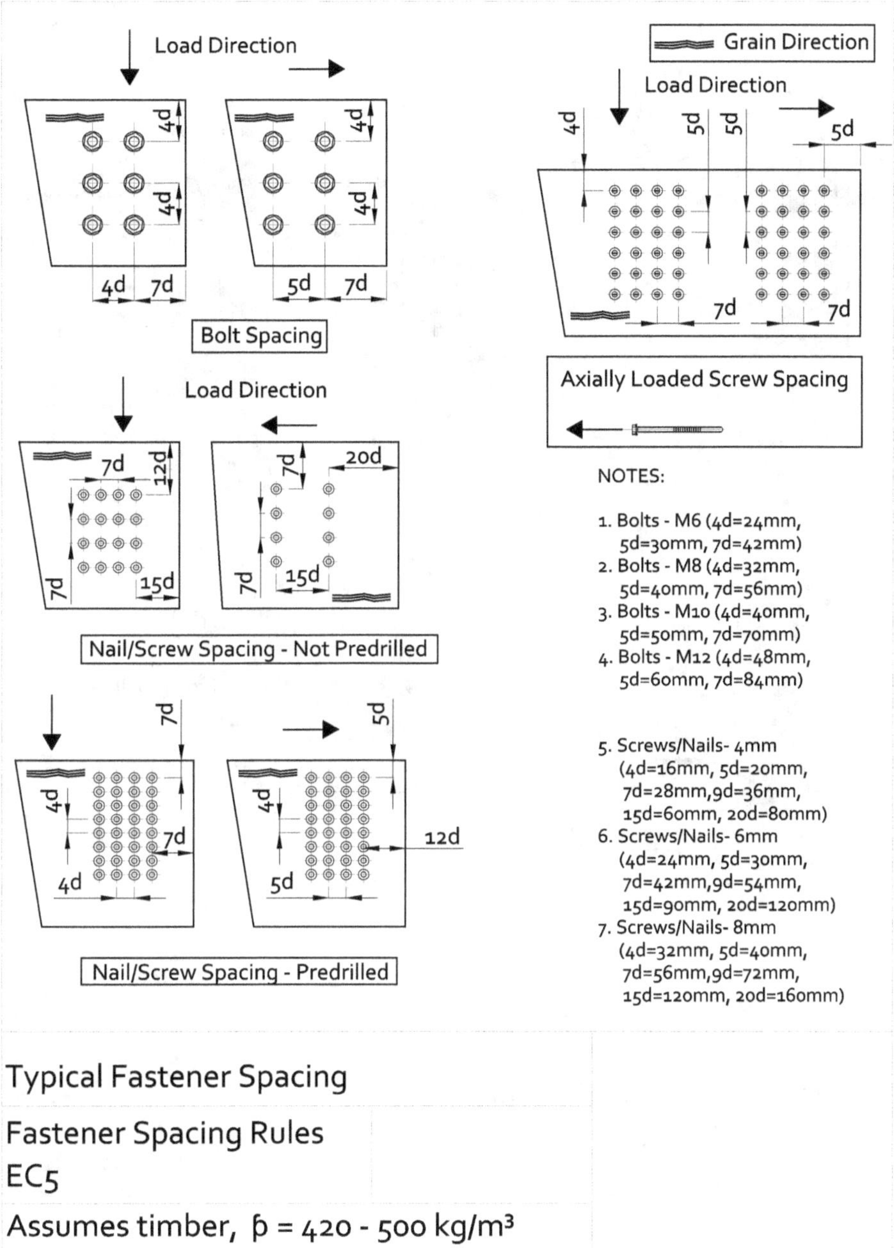

Figure 8.3.1 Screw and Bolt Fixings Spacing and Edge Distances

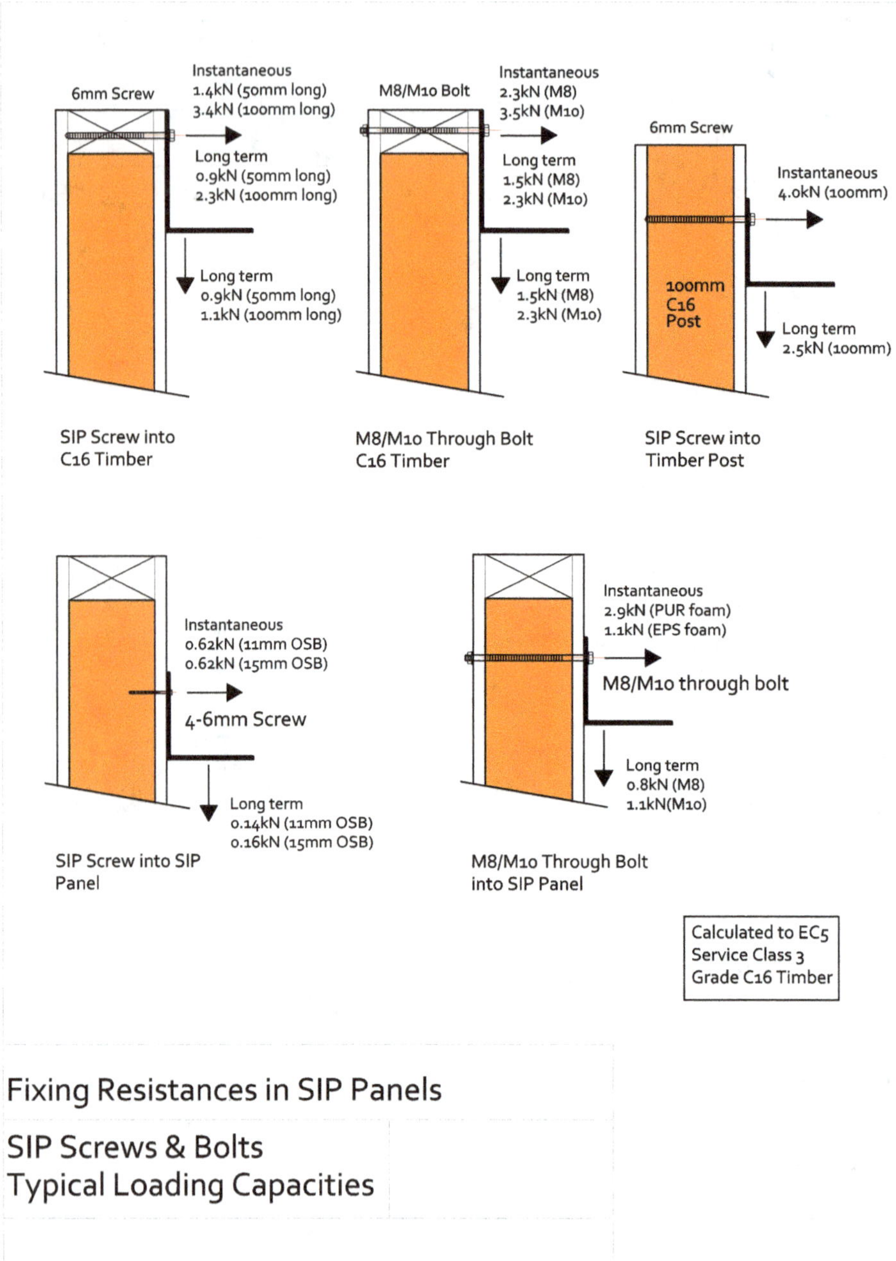

Figure 8.3.2 Screw and Bolt Fixings Typical Resistances For Single Fixings

Relevant Sections

Chapter 2 (Bending) **Section 2.12.1**
Chapter 6 (Cladding) **Section 6.3**

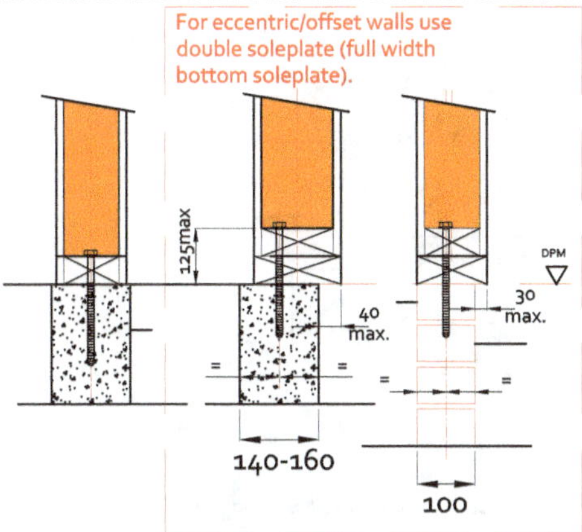

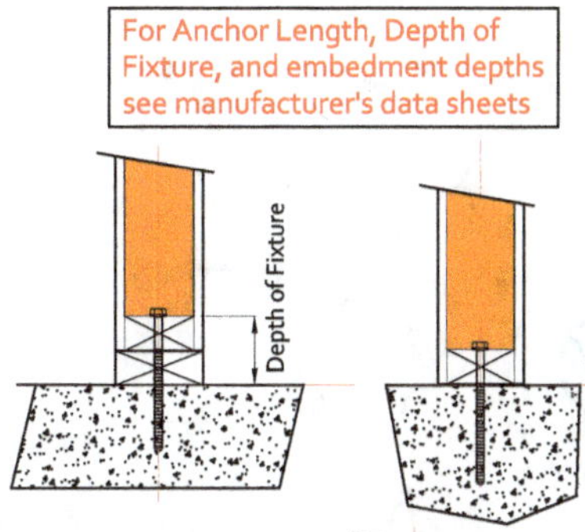

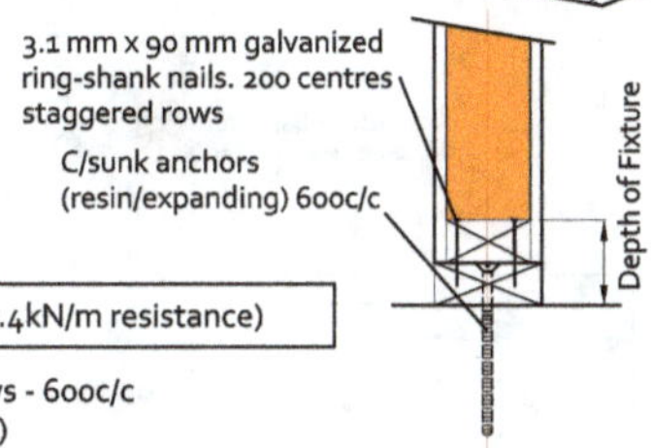

Alternative -Double plate fixing

Sheathed Timber Stud & SIP Walls

Fixing Options (2.0kN/m resistance is approximately unfactored resistance of *single sheathed stud wall)*

7N Blockwork (2.2kN/m resistance)

Thunderbolt M8 - 600 c/c -cap. 3.3kN/m(*1.3kN SWL per anchor*) (also Tapcon 4F screws)

M8 Rawl Loosebolt (expanding anchor) - 500 c/c

M10 Chemfix (Resin Anchor) - 600c/c

M8/M10/M12 Fischer FIS V (Resin Anchor) - 600c/c
- see notes on studs C20-25 Concrete.

C20-25 Concrete (3.4-5.4kN/m resistance)

Thunderbolt M8 screws - 600c/c (also Tapcon 4F screws)

M6 Rawl SPT expanding anchors - 600c/c

M8 Chemfix resin anchors - 600c/c

M8/M10/M12 Fischer FIS V - 600 c/c
Fixture 50mm-M8x175 stud
Fixture 100-110mm - M10x200 stud
Fixture 110-125mm - M12x260 stud

Brickwork (3.2kN/m resistance)

Thunderbolt M8 screws - 400 c/c
(also Tapcon 4F/40)
Exanding anchors not applicable as fixing into 100mm width

M8 Chemfix resin anchor - 600c/c (at 3.4kN/m)

Thermalite (3.5-7 N - 2kN/m resistance)

M10 Chemfix/Fischer FIS V resin anchors - 400c/c

Sole Plate Details

Sole Plate Fixings
Permissible Racking Shear

Loads quoted are 'safe working loads' (serviceability state). They should be checked against available manufacturer's up-to-date data; taking into account anchor length, soleplate depth and embedment.

Figure 8.3.3 Sole Plate Details

Relevant Sections

Chapter 3 (Racking) **Section 3.2.5** and **3.2.6**
Chapter 4 (Compression) **Section 4.2**

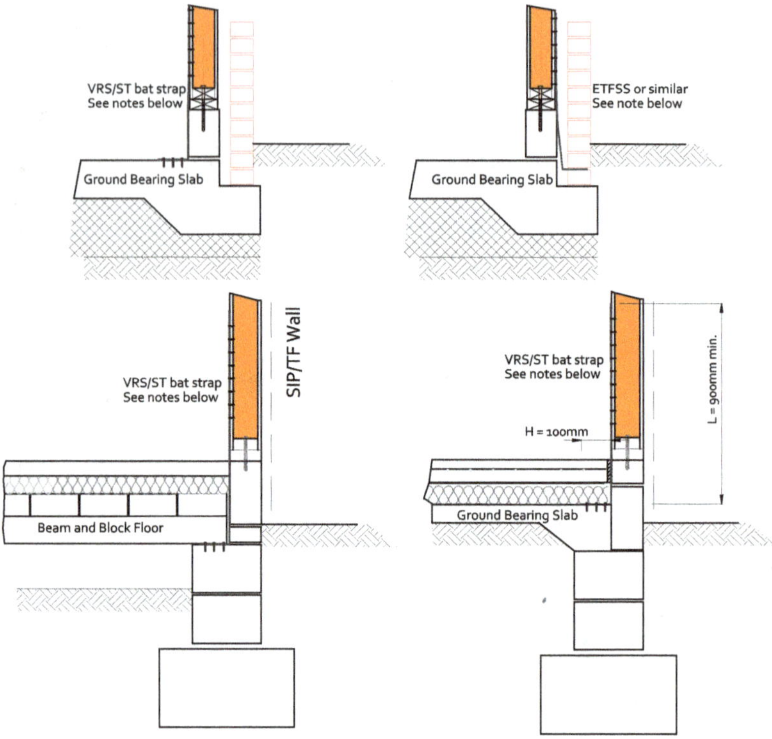

NOTES:

VRS STRAP (ITW/CULLEN)

6No. 4mm screws in SIP/Solid spline vertical leg minimum)

Design capacity in 11/15mm OSB 2.0kN (instantaneous uplift)

Anchors into base to provide capacity of 2.8kN (instantaneous)

NOTES:

Expamet ST BAT Straps

6No. 4mm screws in SIP/Solid spline vertical leg minimum)

Design capacity in 11/15mm OSB 2.0kN (instantaneous uplift)

Anchors into base to provide capacity of 2.8kN (instantaneous)

NOTES:

SIMPSONS ETFSS TIE (ITW Cullen ST-PFS)

Cavity wall tie. Strap to outer masonry skin.

6no. screws into timber/OSB minimum.

Capacity in 11/15mm OSB 2.8kN (instantaneous uplift)

Anchors into base to provide capacity of 2.0kN (instantaneous)

Sole PlateDetails: Strapping Fixings

Sole Plate
Strapping Details

Figure 8.3.4 Sole Plate Details with Additional Straps

Relevant Sections

Chapter 3 (Racking) **Section 3.2.3**
Chapter 3 (Racking) **Section 3.2.5** and **3.2.6**
Chapter 3 (Racking) **Section 3.4.2**

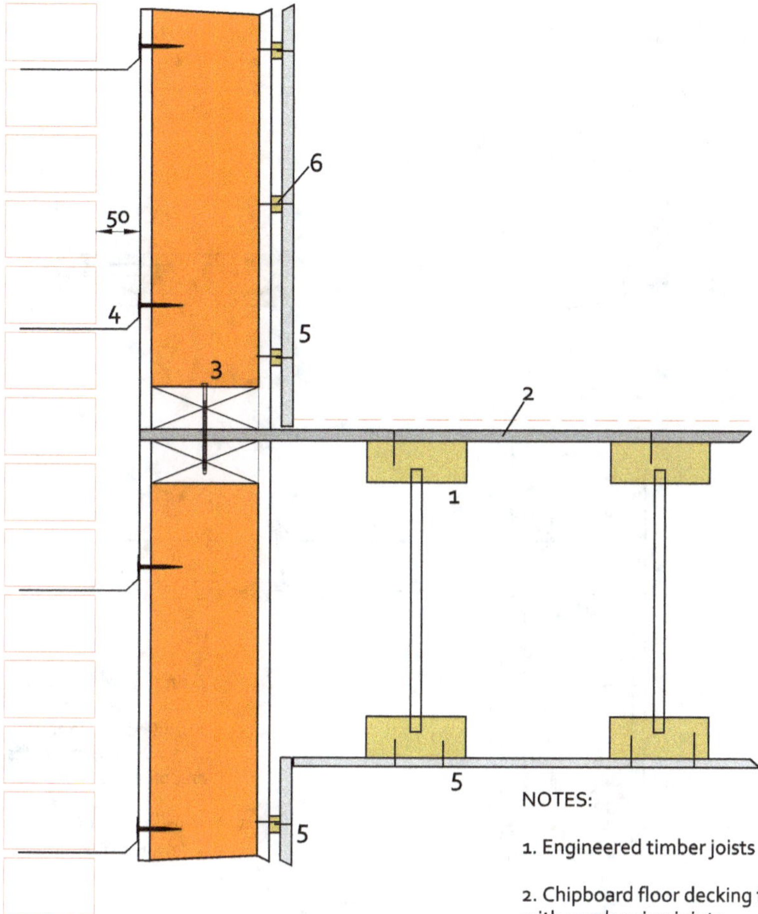

Typical External Wall/Floor Junction

Joists Parallel to Wall

NOTES:

1. Engineered timber joists

2. Chipboard floor decking fitted with overlapping joints

3. 100mm long 6/8mm Headlok screw, EJOT or similar at 350mm centres

4. Brick ties at 450mm vertical centres and 600mm spacing

5. Plasterboard and skim

6. 15mmx35mm battens fixed with 4mm screws at 400mm centres

Figure 8.3.5 External Wall to Floor Junction

Relevant Sections

Chapter 3 (Racking) **Section 3.1.2** & **Section 3.6.1**
Chapter 4 (Compression) **Section 4.4**
Chapter 6 (Cladding) **Section 6.3**

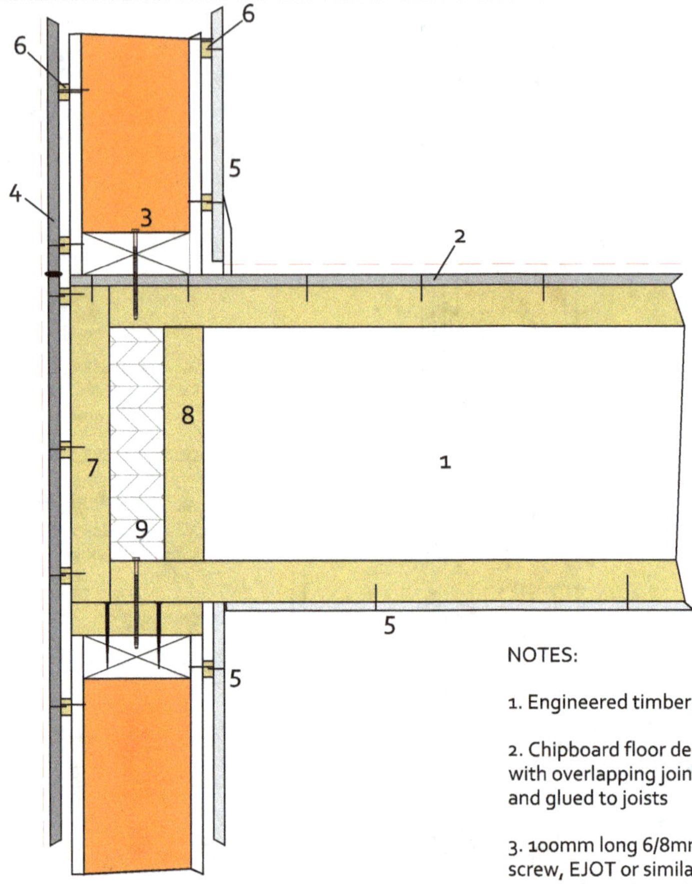

NOTES:

1. Engineered timber joists
2. Chipboard floor decking fitted with overlapping joints, nailed and glued to joists
3. 100mm long 6/8mm Headlok screw, EJOT or similar at 350mm centres fixed through joist/wall
4. Render board and render
5. Plasterboard and skim
6. 15mmx35mm battens fixed with 4mm screws at 400mm centres
7. Continuous rim board
8. Timber blocking
9. Insulation

Typical External Platform Wall/Floor Junction

Joists Perpendicular to Wall

Figure 8.3.6 External Wall to Floor Junction (Platform Frame Type)

Relevant Sections

Chapter 3 (Racking) **Section 3.1.2**
Chapter 6 (Cladding) **Section 6.3**

NOTES:

1. SIPs wall panel

2. SIPs pitched roof panel

3. 250mm long 6/8mm Headlok screws, EJOT or similar at 350mm spacing

4. Profile cut continuous timber nailed and glued to header

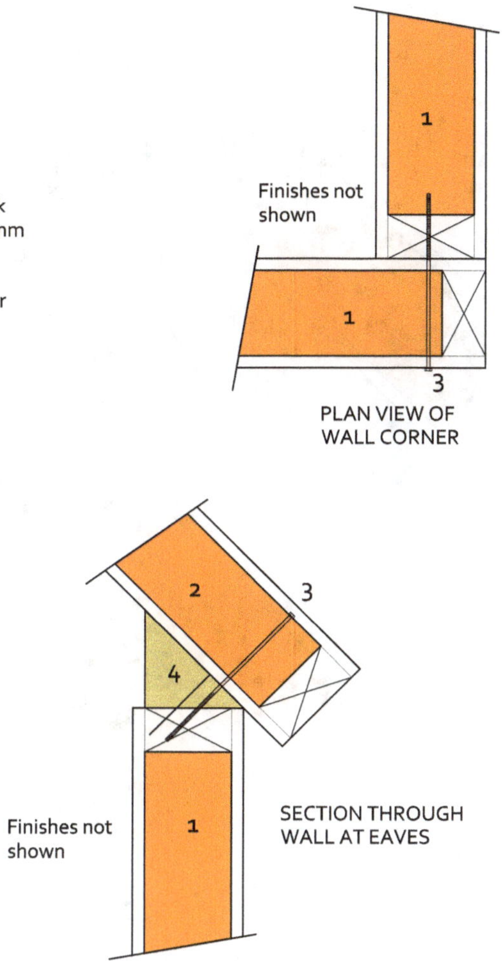

Typical External Wall Corner and Eaves Details

Figure 8.3.7 Typical Eaves Wall Corner and Eaves Detail

Relevant Sections

Chapter 2 (Bending) **Section 2.12.1**

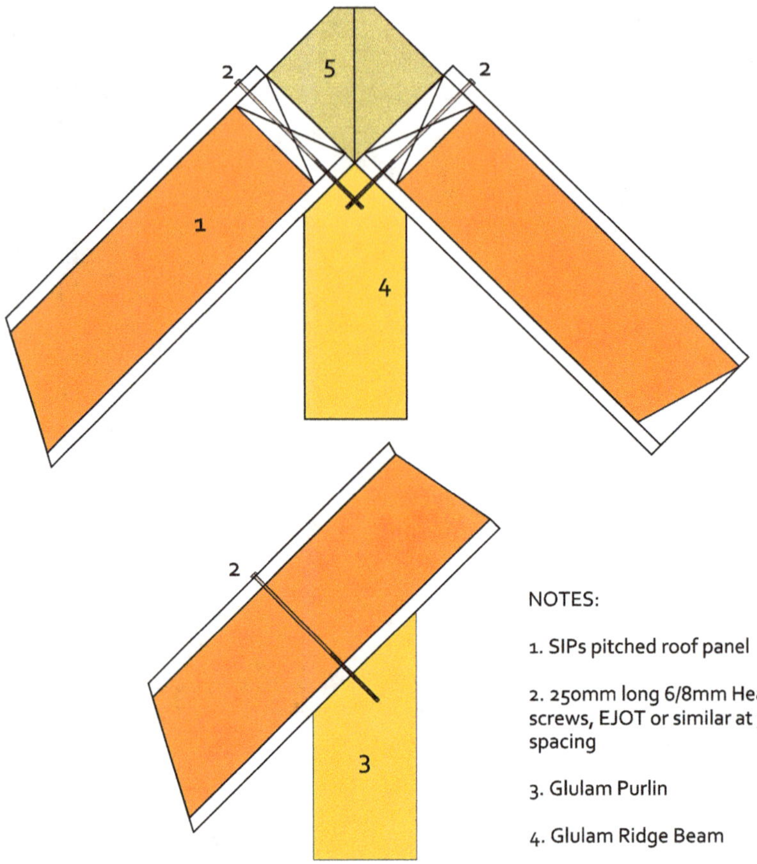

NOTES:

1. SIPs pitched roof panel

2. 250mm long 6/8mm Headlok screws, EJOT or similar at 350mm spacing

3. Glulam Purlin

4. Glulam Ridge Beam

5. Profile cut continuous timber nailed and glued to header

Typical Ridge and Purlin Details

Figure 8.3.8 Typical Ridge and Purlin Details

Relevant Sections

Chapter 2 (Bending) **Section 2.12.1**

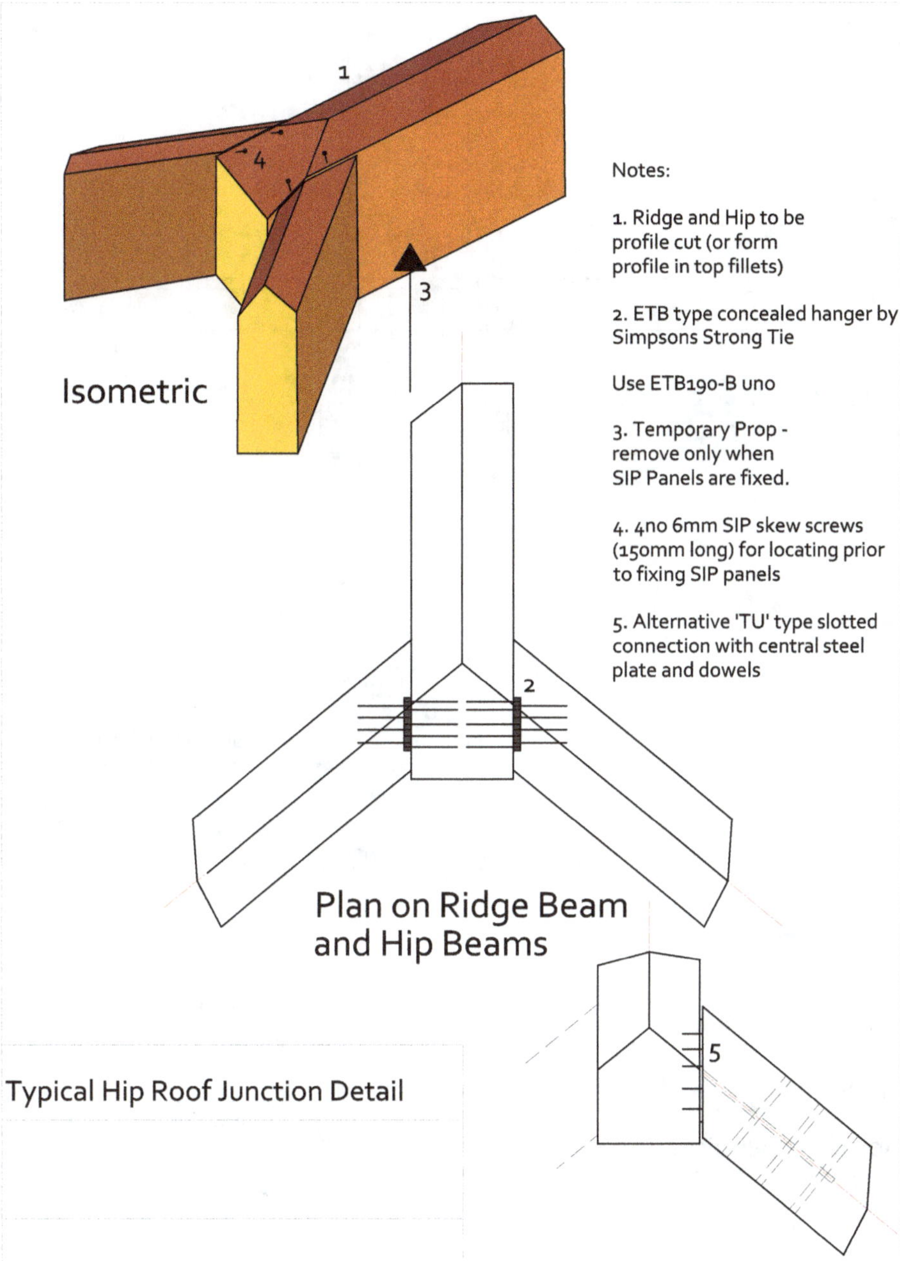

Figure 8.3.9 Typical Hip Roof Junction Detail

Relevant Sections

Chapter 1 (Introduction) **Figure 1.3.20 and 1.3.21**

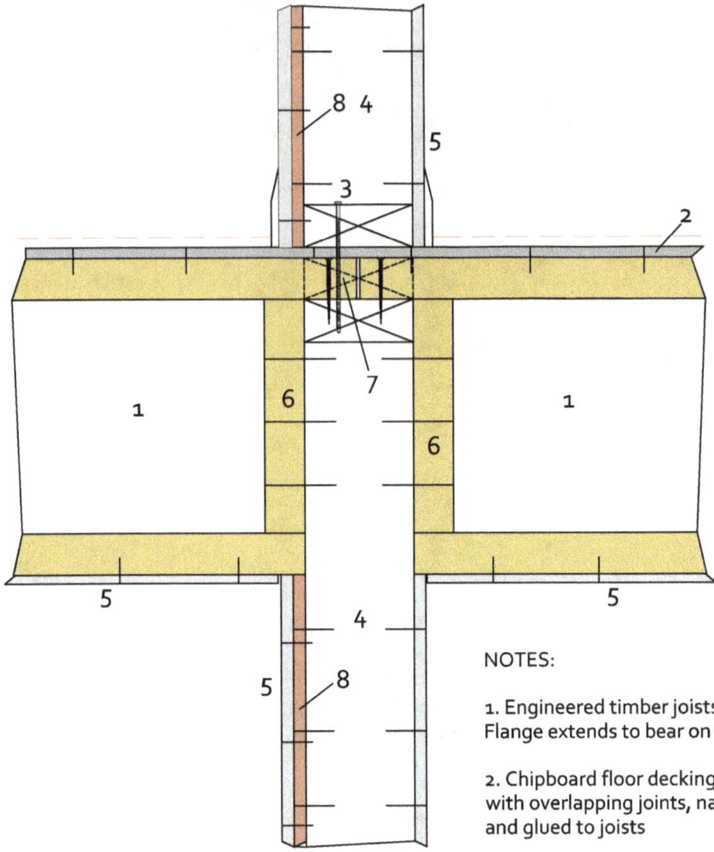

NOTES:

1. Engineered timber joists. Flange extends to bear on wall.

2. Chipboard floor decking fitted with overlapping joints, nailed and glued to joists

3. 250mm long 6/8mm Headlok screw, EJOT or similar at 350mm centres fixed through joist/wall

4. Stud Wall

5. Plasterboard and skim

6. Timber blocking fixed to studs where applicable

7. Timber blocking between joist overhangs (hidden)

8. OSB sheathing

Internal Load Bearing Sheathed Wall Section

Joists Perpendicular to Wall

Figure 8.3.10 Internal Load Bearing Sheathed Wall Section

Relevant Sections

Chapter 3 (Racking) **Section 3.4**

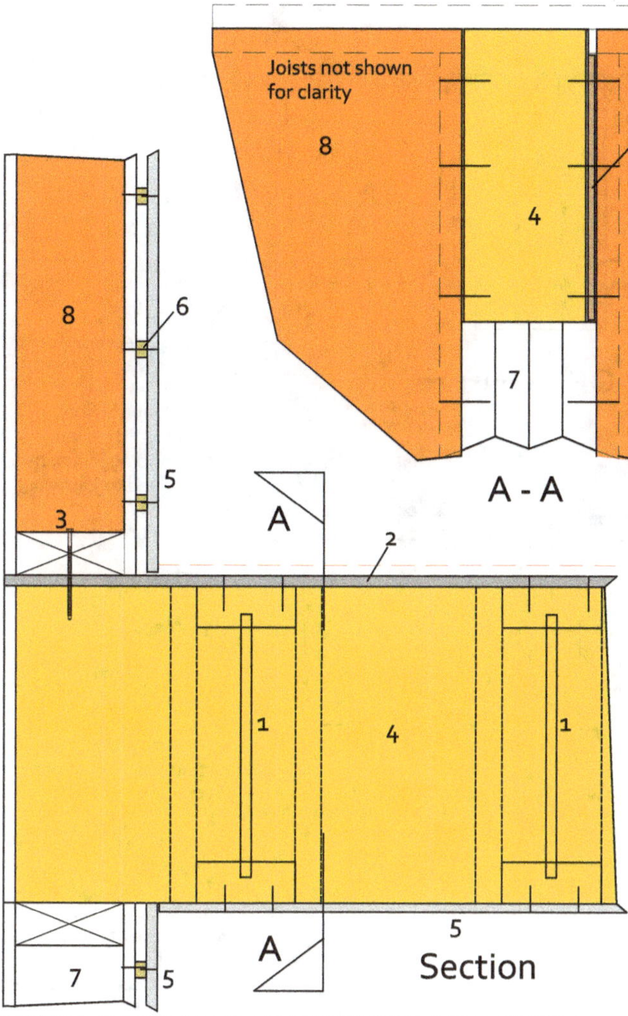

Timber Trimmer to External Wall Connection

Joists Perpendicular to Wall

NOTES:

1. Engineered timber joists. Fixed to glulam trimmer with joist hangers

2. Chipboard floor decking fitted with overlapping joints, nailed and glued to joists

3. 100mm long 6/8mm Headlok screw, EJOT or similar at 350mm centres fixed through joist/wall

4. Glulam trimmer

5. Plasterboard and skim

6. 15x35mm battens

7. Support posts built into external wall

8. SIP external wall fixed to side of trimmer and support posts with 50mm wide edge trimmer

9. OSB packing strip

Figure 8.3.11 Timber Trimmer to External Wall Connection Detail

Relevant Sections

N/A

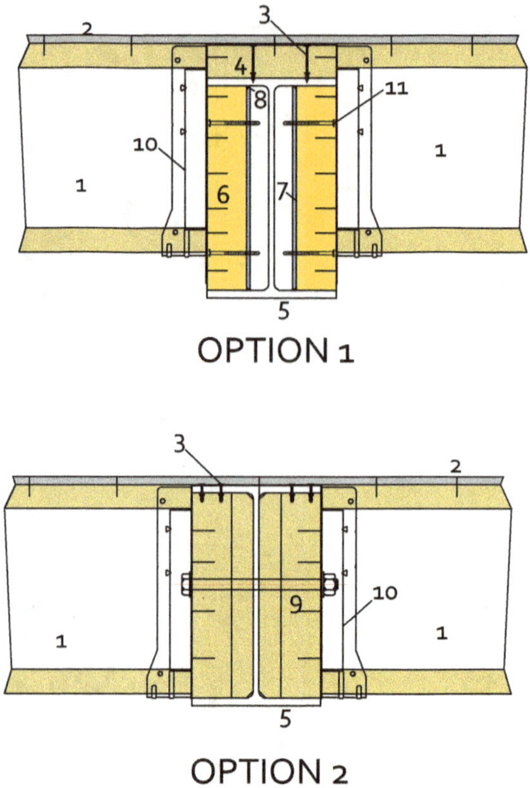

NOTES:

1. Engineered timber joists.

2. Chipboard floor decking fitted with overlapping joints, nailed and glued to joists

3. Shot fired nails

4. Timber plate

5. Steel beam

6. 45mm rim board glulam cut to suit

7. 4mm x 50mm wide steel plates at 600mm centres

8. 4mm prepared FW to flange

9. Timber packing flush with face of flange fixed to web with M8 through bolts at 600mm centres

10. Hanger to suit engineered timber joists

11. 2no. 4/6mm self tapping screws (counterbored/ flush or countersunk) per side fixed into steel plates

Joist to Steel Beam Connection

Joists Perpendicular to Beam

Figure 8.3.12 Joist to Steel Beam Connection

Relevant Sections

N/A

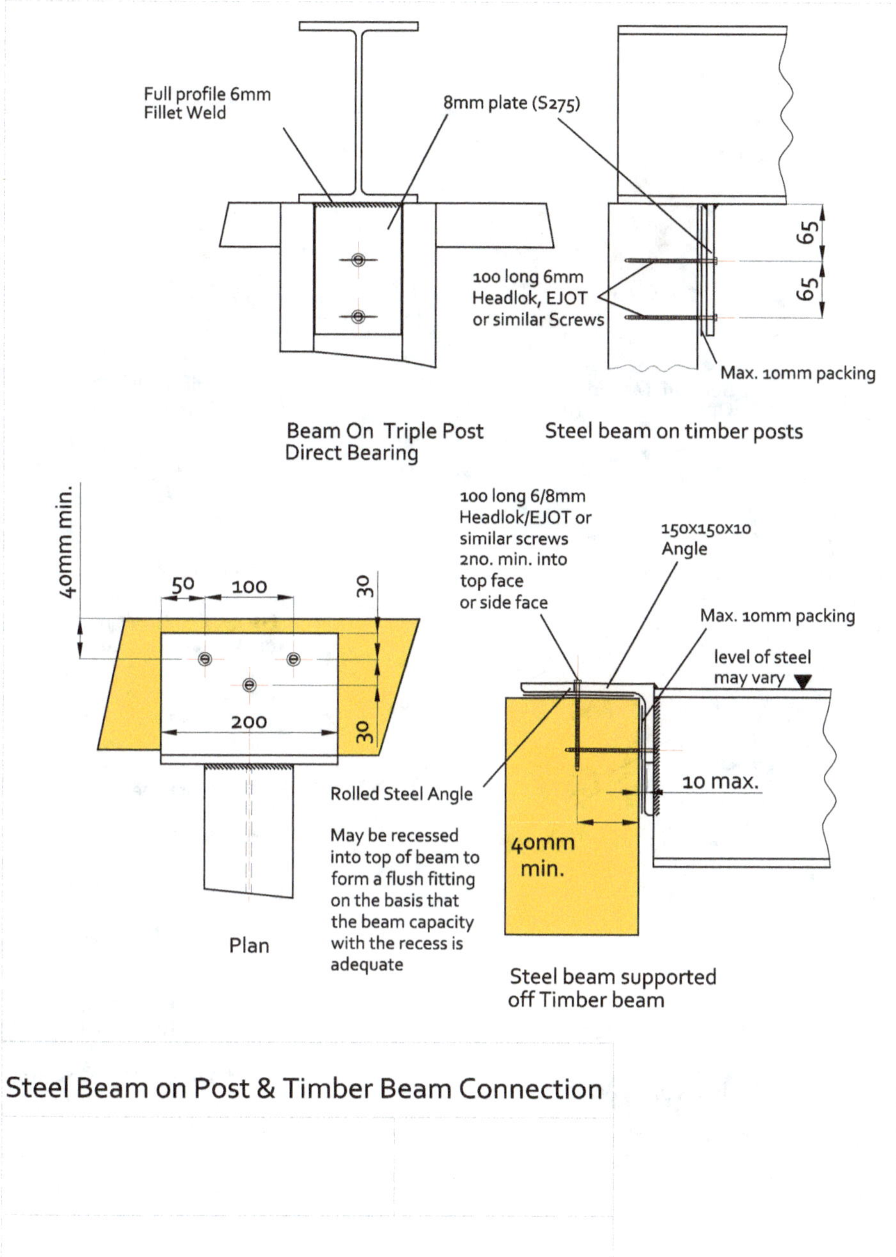

Figure 8.3.13 Steel Beam on Post and Timber Beam Connection

Relevant Sections

N/A

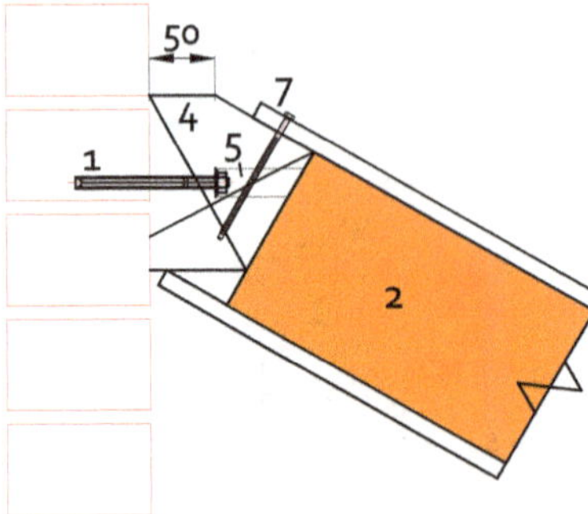

Section (Masonry wall)
Option 1

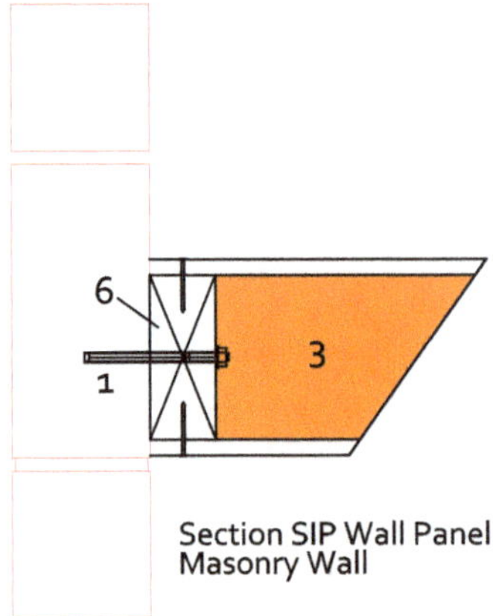

Section SIP Wall Panel
Masonry Wall

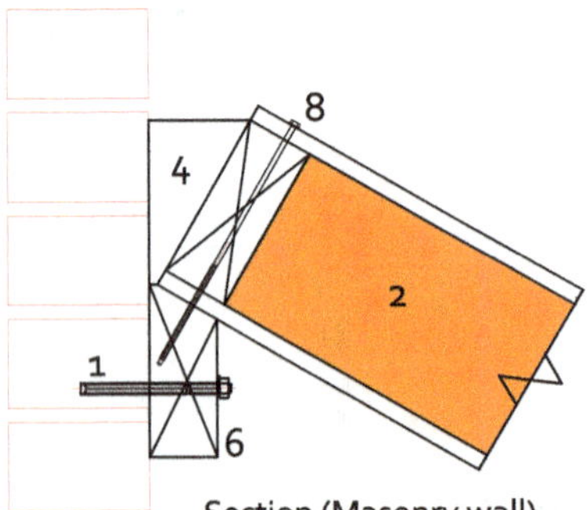

Section (Masonry wall)
Option 2

NOTES

1. M8mm Chemfix (resin anchors) or 4 No M8 HD Bolts: Fischer FIS VT (resin anchors) or M10-60L Rawl Loosebolt expanding anchors. Ensure anchors have correct embedment including for 50mm fixed timber.

2. SIP Roof Panel

3. SIP Wall Panel

4. C16 (Profiled) Timber

5. Counterbored to 50mm from wall face

6. 50x125 Bearer (C16) chamfered to suit roof slope

7. 6/8mm 150mm long screws Headlok/EJOT or similar at 300mm centres

8. 6/8mm 200mm long screws Headlok/EJOT or similar 300mm centres. Minimum timber embedment 38mm

SIP Panels to Masonry Wall Connection

Timber Plate Connection

Figure 8.3.14 SIPs Roof Panels to Masonry Wall Connection (Timber Plate)

Relevant Sections

N/A

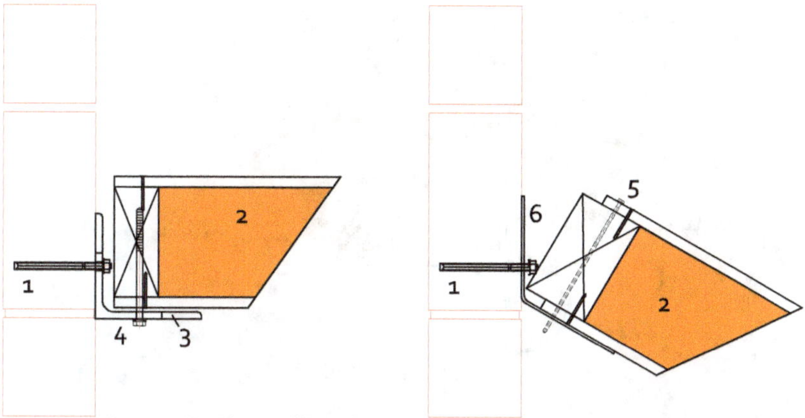

Section SIP Flat Roof Panel Masonry Wall

Section SIP Sloping Roof Panel Masonry Wall

1. M8mm Chemfix (resin anchors) or 4 No M8 HD Bolts - Fischer FIS VT (resin anchors) or M10-60L Rawl Loosebolt expanding anchors

2. SIP Roof Panel

3. 120x120x8 RSA or bent 3mm angle (6)

4. 150 Long SIP Screw/Headlok @ 350mm c/c

Or (5) self tapping screws from top of panel into centre of angle

5. Self tapping screws from top of panel into centre of angle

6. 3mm angle, bent to form 120x120mm legs. (from 240mm wide plate)

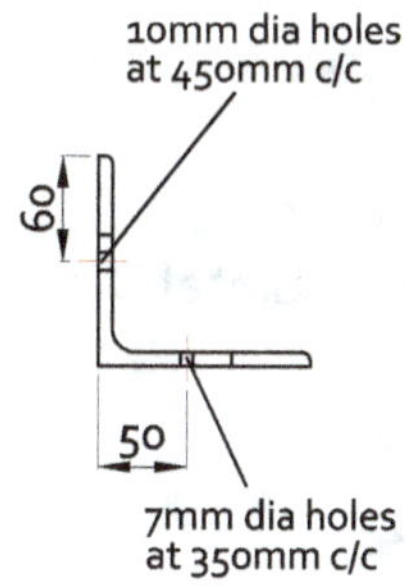

SIPs Roof Panels to Masonry Wall Connection

Steel Angle Connection

Figure 8.3.15 SIPs Roof Panels to Masonry Wall Connection (Steel Angle)

Relevant Sections

N/A

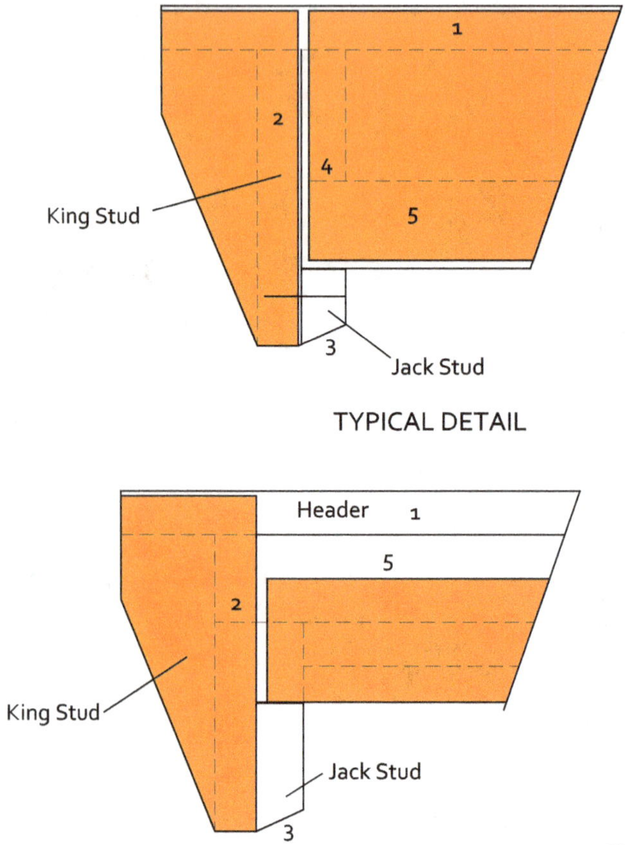

ALTERNATIVE (SHALLOW DEPTH)

notes:

1. 50mm header plate

2. Continuous 50mm king stud

3. Jack stud 50mm - nailed and glued at 300mm centres with 4mmx100 nails into adjacent post

4. 50mm cripple stud in end recess

5. Lintel - as specified

Typical Lintel Details at External Wall

Steel Angle Connection

Figure 8.3.16 Typical Lintel Details at External Wall

Relevant Sections

Chapter 5 (SIPs as Beams) **Section 5.1**

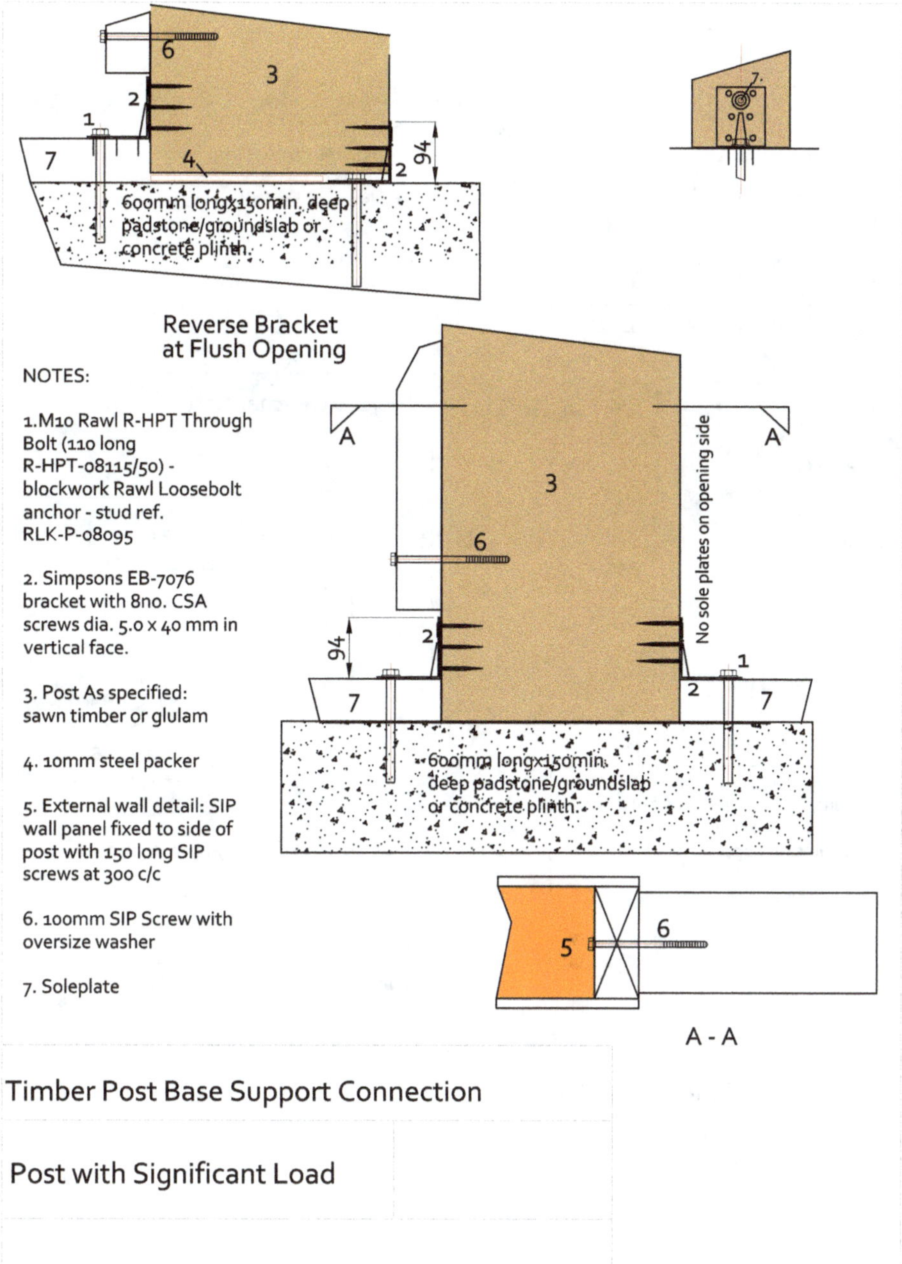

Figure 8.3.17 Timber Post Base Connection with Significant Load

Relevant Sections

N/A

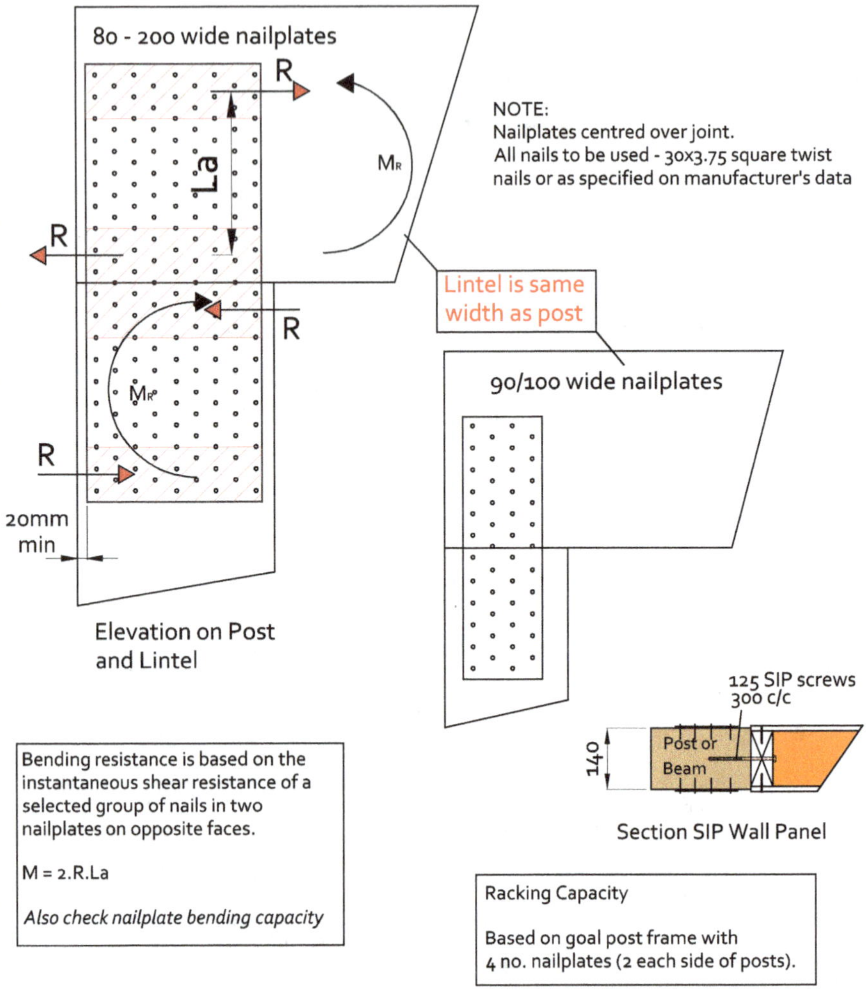

Nailplate Racking Frame Detail

Connection at Head of Post

Figure 8.3.18 Nailplate Connection at Head of Racking Frame

Relevant Sections

Chapter 1 (Introduction) **Section 1.3.6 (Figure 1.3.17)**
Chapter 3 (Racking) **Section 3.5 (Figure 3.5.3)**

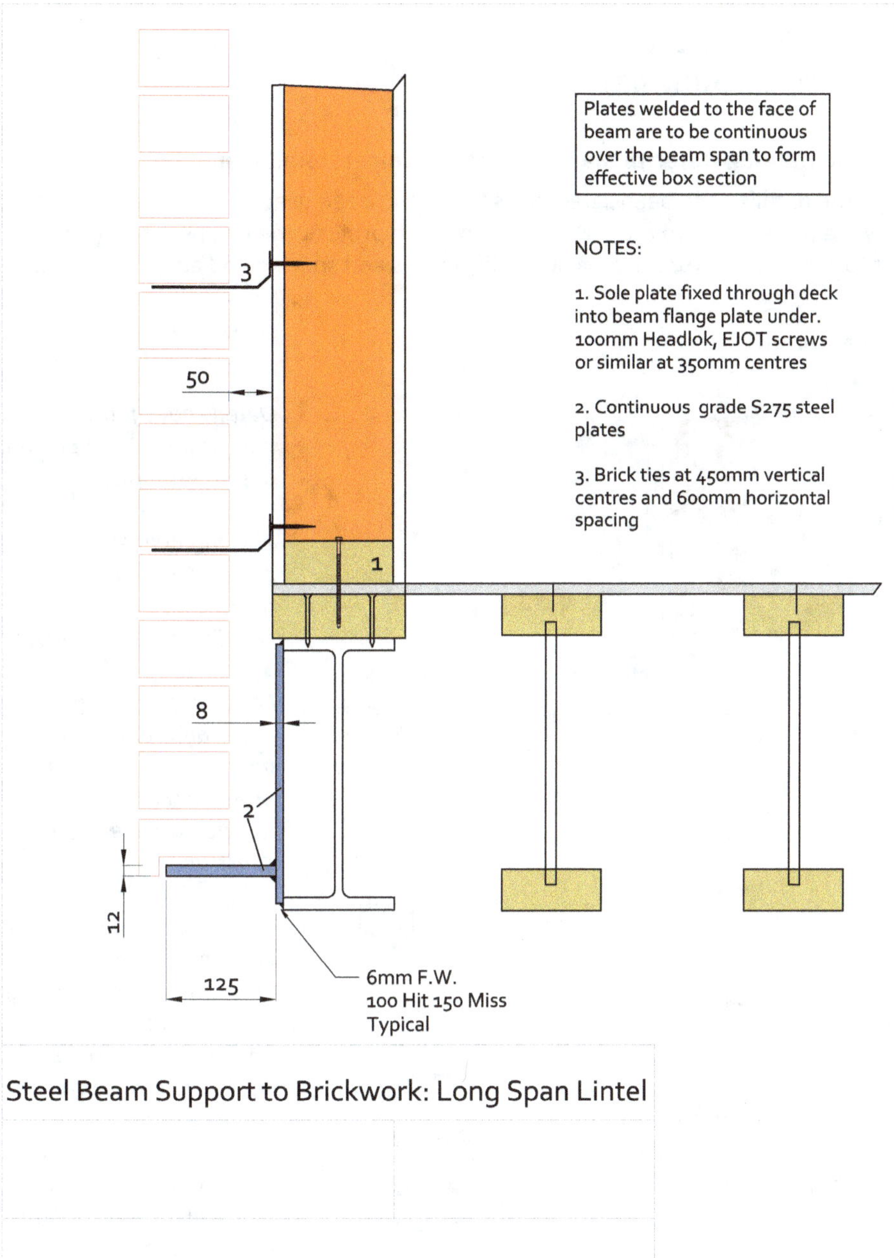

Figure 8.3.19 Steel Beam Support to Masonry as Long Span Lintel

Relevant Sections

N/A

Chapter 8

Quick Note Summary

Generic Details: Refer to Manufacturer's Typical Detail Manuals

All of the details have been selected as they have been used in a number of projects. They are not specific to the panel type. For each project, the engineer should refer first to the manufacturer's typical details, fixings and spacing and adapt their details to suit.

Typical Detail

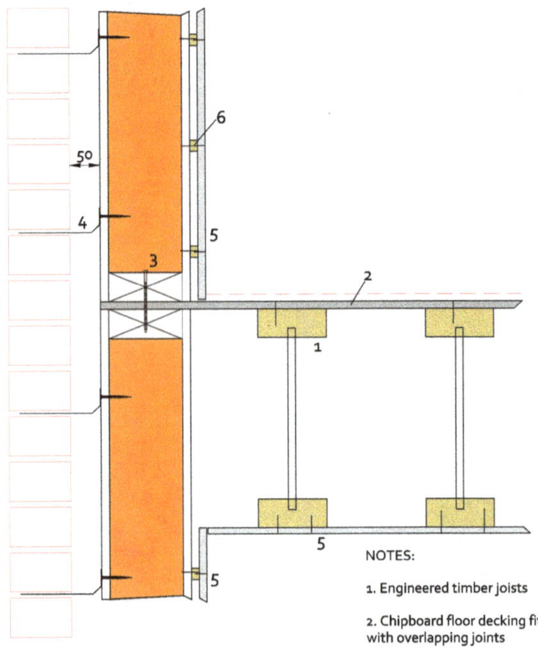

Typical External Wall/Floor Junction

Joists Parallel to Wall

NOTES:

1. Engineered timber joists
2. Chipboard floor decking fitted with overlapping joints
3. 100mm long 6/8mm Headlok screw, EJOT or similar at 350mm centres
4. Brick ties at 450mm vertical centres and 600mm spacing
5. Plasterboard and skim
6. 15mmx35mm battens fixed with 4mm screws at 400mm centres

Details have footnotes that highlight relevant previous design examples.

The engineer should research data for fixings and their capacities if the supplier has alternatives.

For example, there are many different screw manufacturers which produce screws produced for the purpose of SIPs construction.

The engineer should check with their client team or SIPs contractor supplying the SIPs construction pack which components are to be used. For instance decking will be supplied and installed with fixings and adhesive generally advised by the manufacturer.

Figure 8.3.5 External Wall to Floor Junction

Relevant Sections

Chapter 3 (Racking) **Section 3.1.2**
Chapter 4 (Compression) **Section 4.4**
Chapter 6 (Cladding) **Section 6.3**

Chapter 9

The Future of SIPs

Further testing for creep, racking, two way spans and skin fixings. Composite action potential for roof purlin interaction with panels. Hybrid structures with SIPs and cross laminated timber (CLT). Possibilities of wider adhesive use. Modular construction - development of social housing types and multistorey units.

9.1 FURTHER GENERIC TESTING FOR MORE ECONOMIC STRUCTURES

In the UK, design guidance and test data for SIPs design are not readily available for engineers. References to testing of SIPs panels in bending, racking and for screw pullout and shear have been added to the relevant chapters, which may be downloaded online. In order to design more economic structures standardised tests are proposed so that each manufacturer can provide consistent and comparable information for the practising engineer. An industry standard for properties, such as core shear modulus, in a similar way to graded timber, would be a great advantage for the SIPs industry. A summary of the proposed testing follows.

Long Term Creep Deflection Tests

Creep testing on all OSB and MgO panel types (142mm Kingspan for example).

Panel spans: 3m for 125-150mm range, and 4m for 150mm and above.
Panel composition: SIP splines and manufacturer standard timber splines.
Panel width: 3 panel width ideal (3.6m) for the best representation of shear lag.

1. Tests with the loading of 0.75kN/m^2 for 12+ weeks, modelling flat roof imposed loads.

2. Tests with the loading of 1.5kN/m^2 for 12+ weeks, modelling higher loads, such as green roofs.

Centre span deflection to be taken initially at half day intervals for a week and then weekly intervals. Representative humidity should be maintained.

A SIP panel acts as a composite of foam and OSB outer sheets. Three locations of this should be marked, and regular recorded thickness measurements taken or strain gauges set into the face of the foam and top of the face to obtain creep strain data to assist with panel evaluation. Representative humidity should be maintained.

Load Levels with Relation to Creep Factors (Kdef) Further Research

Long term research at Eindhoven University of Technology[18] was conducted over a period of 12 years on panels with 200mm depth EPS foam core and with thin steel faces. It was found that, for panels loaded where the applied bending was circa 40% or less of the characteristic bending strength of the panels, the creep factor for EPS foam, k_{def}, would never be larger than 1.6 after 50 years. For the roof panel example in **section 2.3.5,** the design bending moment was circa 15% of the *characteristic* strength. Assuming K_{def} = 1.9, this gave a permanent sag of 20mm (half the sag for k_{def} = 7 in **table 2.3.1**). There is scope, from these encouraging results, to reduce creep factors considerably for typical permanent load levels on roofs and floors.

Compression Testing on Panels (Unreinforced and Reinforced)

Compression Tests on Panels. 2.4m wide. SIP Splines.

1. Panels at 3m height
2. Panels at 6m height.

Compression Tests on Panels 2.4m wide. Timber Splines. (Reinforced)

1. Panels at 3m height
2. Panels at 6m height.

Racking tests

Racking Tests on OSB and MgO panel types. 2.4m wide. SIP & Timber splines.

1. Racking test on 2.4x2.4m panel
2. Racking test on 3.5m high panel (unreinforced) - *optional*
3. Racking test on unreinforced 6m high panel - *optional*

All fixings, adhesives or silicone sealants to be applied representative of the SIP system building standard details.

Shear Testing on Foam Samples

Establish the shear modulus for the different cores used (EPS/XPS/Polyurethane) using standardised methods.

Two Way Span Tests

It is known that SIP panels laid out on a roof will exhibit two way spanning action. Calibrated tests comparing panels with 2 edge and 4 edge support with SIP and timber spline options would be beneficial for finite element modelling parameters, as discussed in **section 2.8**. It would not necessarily have to be long term loading.

Many pitched or flat roofs have battens fixed perpendicular to panel layouts to the top for tiles or plywood decking and sometimes for ceiling plasterboard installation. This build-up would contribute to 2 way spanning rigidity. Additional tests to compare results with tested panels without battens would give improved confidence in models.

Standard Pullout and Shear Tests on Screw and Nail Face Fixings

All MgO and OSB panels should have standardised pullout and shear tests to verify fixings recommended by each manufacturer.

9.2 THE POTENTIAL FOR COMPOSITE ACTION BETWEEN PANEL AND BEAM

In **figure 9.2.1,** a roof panel layout is supported on intermediate glulam purlins. It is posited that the interaction of beam and panel act in a composite manner. An investigation into a 142mm panel with idealised short term modulus (see **section 2.7**), assuming 200mm long SIP screws fixed to a 315x115 glulam purlin, was calculated with formulae derived from standard methods for steel with composite deck. With SIP flange width of 1.5m, allowing for shear lag, and effective 50% composite action gave a four fold increase in the purlin transformed stiffness. Even accounting for long term loading and reduced stiffness perpendicular to the panels, this indicates potential for further research.

Purlin supporting roof panels with regular screw fixings providing potential composite action.

Splines not shown for clarity.

Figure 9.2.1 Roof Beam With Regular Screwed Fixings

9.3 HYBRID CROSS LAMINATED TIMBER (CLT) AND SIPS CONSTRUCTION

SIPs construction is largely limited to three storey construction. Cross laminated timber panels made from glue laminated layers have been adopted for larger scale projects such as schools or multistorey apartments.

CLT floors can be constructed from solid panels up to circa 320mm deep, and spanning up to 6m or greater. CLT walls have greater racking capacities than SIPs walls, typically up to 15kN/m; compared with 2.5-5.6kN/m for SIPs walls.
See **section 3.2.1**.

There is no reason why the advantages of SIPs insulative and airtightness properties cannot be realised by combining CLT floors and shear walls, for instance, with SIPs external walls for 'all timber' projects, such as schools or apartments.

9.4 ADHESIVE MOMENT RESISTING JOINTS

There have been advances in adhesive technology and a range of weather resistant adhesives, such as resorcinol formaldehyde (finger jointing and laminating) or one part polyurethane. A moment resisting joint is shown in **Figure 9.4.1**.

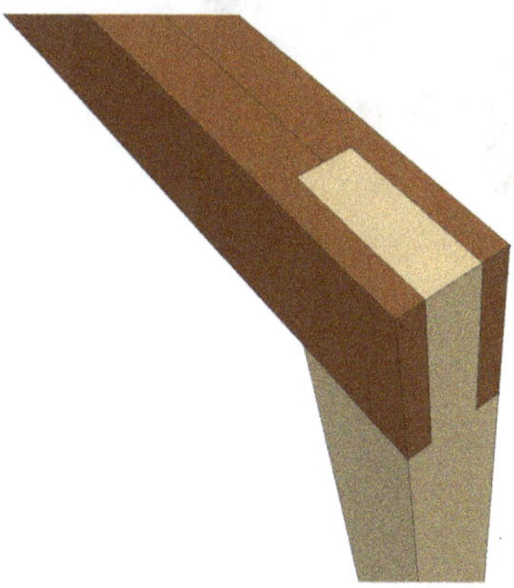

Figure 9.4.1 Adhesive Bonded Knee Joint

For many commonly used timber adhesives, a properly applied adhesive to a clean joint offsite may be stronger than the shear bond in the timber fibres. For low rise modular construction, it should be possible to form robust, long life moment resisting joints from adhesive alone, which is equivalent to the moment resistance of the timber components. Each joint could be proof tested prior to delivery.

In **figure 7.1.1** portalised frames with adhesive bonded joints were proposed, avoiding the need for internal steel splice plates, thereby creating a timber structure in its entirety, with improved thermal performance.

9.5 WIDER ADOPTION OF MODULAR CONSTRUCTION

In **chapter 7** modular construction was discussed with SIPs modular units being demonstrated as well suited to modular construction, with great benefits from its racking strength and insulative properties.

Figure 9.5.1 shows the stacking of units, factory assembled, to form a three storey building such as the one illustrated in **figure 7.2.2**. With a worldwide need for fast track buildings for dwellings this unarguably will have more widespread adoption in the future.

For terraced housing, commonly built for social housing, it is possible to combine SIPs with portal timber frames in LVL type or glulam with internal steel splines to form fast erected and repetition to reduce site and offsite costs.

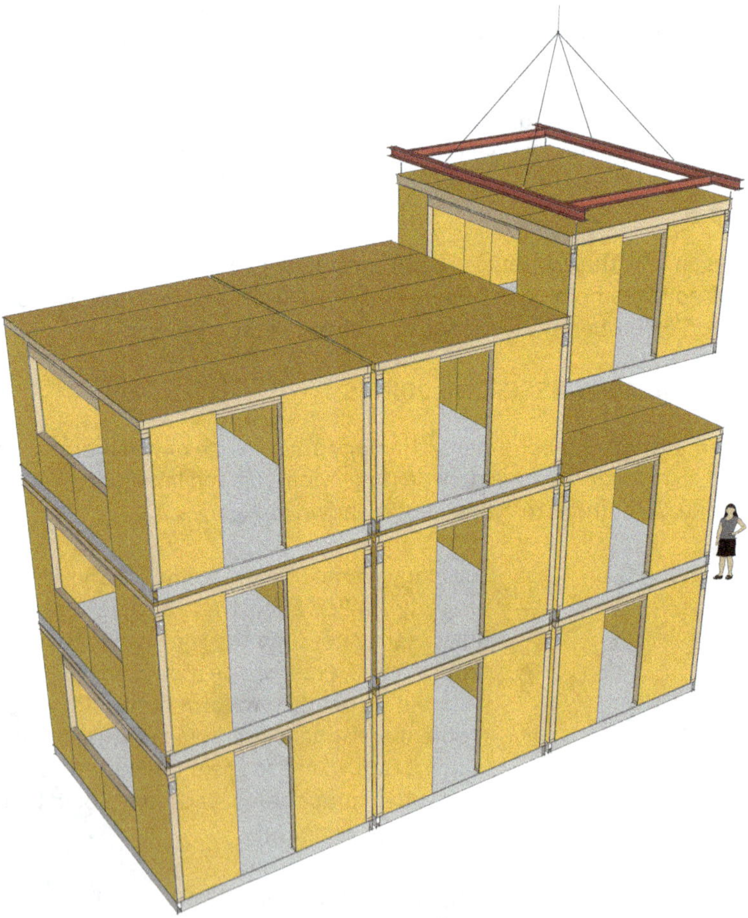

Figure 9.5.1 Multistorey Stacked Modular Units

Glossary of Terms

anchor	Refers to fixings of ground level soleplates to supporting structure.
batten	Treated timber sections fixed to external SIP face in rows at set spacing to support cladding: tiles, brick slips, render boards etc.
cavity	Space between wall skins or between wall and cladding, which may or may not be filled with insulation.
bracing	Stabilising framing or sheathing to restrain a wall, floor or roof.
sill plate	Timber section at the base of a window opening.
DPM	Damp proof membrane: waterproof barrier at ground level.
face	Outer skin or face of a SIP panel - most frequently OSB.
flexural	Describes action of bending.
flexure	Bending: 'bowing' or sagging/hogging of structures under. transverse or 'out of plane' (perpendicular) loads.
hanger	Proprietary bracket fixing a joist end to its support.
head binder	See header.
header	Timber section fixed into top horizontal wall recess.
improved nail	Nail with ribbed shaft for increased withdrawal resistance.
jack post	Opening post at the edge lintel bearing support.
jack stud	See jack post.
jamb	Side post or vertical face of wall opening.
king post	Opening inner jamb post to full height of wall panel.
king stud	See king post.
lintel	Support to structure or fabric at the top of an opening in a wall.
OSB	Orientated Strand Board - generally grade 3 for environments where there is greater humidity, suited to construction and house building.
overhang	Part of a roof or floor extending past the exterior wall.
plate	Any timber section fitted to panel recesses or to a wall or beam.
purlin	Intermediate roof panel support beam.
racking	Generally referring to walls: structural resistance to lateral loads.
raking	Refers to leaning or sloping structural elements.
sheathing	Usually OSB or plywood sheets fixed to a stud wall to form bracing.
SIP screw	Wide headed wood screws up to 350mm long. Examples: ESCR washer head screw or Fastenmaster Headlok.
SIPs	Structurally Insulated Panels.
soleplate	Single or double timber treated section at the base of a wall (at ground floor level or any upper storey).
spline	SIP or timber which forms connection between SIP panels.
stud	Vertical timber post arranged in a line of studs forming a stud wall.
stud wall	Wall comprising studs with or without sheathing.
rolling shear	Shear stress or line load in a plane perpendicular to the grain direction.
shear flow	'Flow' of shear stress or line load internally in a beam or wall resulting from transverse loads in a beam or lateral loads in a wall. Interchangeable with *rolling shear* in a beam or wall spanning over an opening.

References

1. Morrison H. "Structural Engineering Art and Approximation", (2019) Paragon, Chapter 12. *Section:* ***2.3, 2.3.2.1***

2. Allen H.G. , "Analysis and Design of Structural Sandwich Panels". (1969) Pergamon Press. *Section:* ***2.3, 2.3.6, 4.1, 4.1.1, 4.3, 4.6***, Chp. **4 summary**

3. Creep (foam) Taylor S.B, Manbeck H.B., Janowiak J.J., Hiltunum D.R. "Modelling Structurally Insulated Panel (SIP) Flexural Creep Deflection." (December 1997) J. Structural Engineering, volume 123, No. 12, . *Section:* ***2.3***

4. Davies J.M. "Design Criteria for Structural Sandwich Panels". (1987) The Structural Engineer, 65A, pp. 435-441. *Section:* ***2.3***

5. European Assessment Document - EAD 140022-00-0304. TR019. *Section:* ***2.3, 2.3.1, 2.3.6, 2.4.1, 2.12***

6. Rungthonkit P, "Structural Behaviour of Structural Insulated Panels". University of Birmingham, PHD Thesis (2012). *Section:* ***2.3, 2.3.1, 2.3.6, 2.11, 2.12, 4.1.1, 4.3, 4.5, 4.6***

7. Stanhope Position Paper (report by Arup): "An Assessment of the Imposed Loading needs for current Commercial Office Buildings in Great Britain." (1992) *Section:* ***2.3.2.2***

8. Tompos E, "IM 14 TIP 01 SIP Design Guide 2009-03-19".(2009). National Timber Association (USA). *Section* ***2.11***

9. Jorissen prof. dr. ir. A.J.M. (TU/e) Hofmeyer dr. ir. H. (TU/e) ing. van Rie J.L.G. (Kingspan Unidek), "Verstijvers in sandwichelementen en spanningsafhankelijk kruipgedrag van EPS" (Stiffeners in Sandwich Elements and Stress-Dependent Creep Behaviour of EPS) (2016), Eindhoven University of Technology. *Section:* ***2.4.1***

10. van Rie J. and Jorissen A. "Skh-publicatie 09-01.praktische rekenmethode voor sandwich en sandwich rib elementen". (Practical Calculation Method for Sandwich and Sandwich Rib Elements). (2009) Technical report, SKH, Wageningen. *Section:* ***2.4.1***

11. Dhonju Racking R.L.,"Performance of Platform Timber Framed Walls." (2018) Edinburgh Napier University and Glasgow Caledonian University, PhD Thesis with test reports. *Section* ***3.1.1***

12. Kuai L., Ormarsson S., Vessby J., Maharjan R.,"A numerical and experimental investigation of non-linear deformation behaviours in light-frame timber walls."Linnaeus University and Karlstad University. *Section* ***3.1.1***

13. TRADA. Timber frame housing: UK Structural Recommendations. *Section* ***3.2.3***

14. Warren.C,"Roark's Formulas for Stress and Strain." (1989) McGraw-Hill, Chp. 11 *Section* ***4.1.1, 4.4***

15. Jorissen A., Van Rie J., Modderman J. Hofmeyer H., " Shear Creep of Polystyrene Cores in Wood-Based Panels", (2023) Eindhoven University of Technology. World Conference on Timber Engineering, Oslo. *Section* ***2.3.1***, ***9.1***

Photographs

In addition to photographs by the writer, clients and other parties have kindly given their permission to include their own images in the book. It is to be noted that the copyright for the images rests with the original source summarised below.

Use of these images by other parties is assumed to be prohibited without permission from the source, summarised below:

Figure 1.3.13 Barn Conversion Before Completion

Courtesy of SI Hus Ltd.

Figure 1.3.14 Barn Conversion After Completion

Courtesy of SI Hus Ltd.

Figure 1.3.15 Alpine House Switzerland

Courtesy of Bentley Projects Ltd.

Figure 1.3.16 Pelletron Building Cumbria

Hugh Morrison (author) . SIP Co Ltd project

Figure 1.3.17 House in Cornwall

Courtesy of SI Hus Ltd.

Figure 1.3.18 School in Cornwall

Courtesy of SI Hus Ltd.

Figure 1.3.19 House Interior - Open Plan Space

Courtesy of SI Hus Ltd.

Figure 1.3.20 Hipped Roof Underside

Courtesy of SI Hus Ltd.

Figure 1.3.21 Hipped Roof Framing

Courtesy of SI Hus Ltd.

Figure 1.4.13 House SIPs First Floor Construction

Hugh Morrison (author). SIP Co Ltd project

Photographs (continued)

In addition to photographs by the writer, clients and other parties have kindly given their permission to include their own images in the book. It is to be noted that the copyright for the images rests with the original source summarised below.

Use of these images by other parties is assumed to be prohibited without permission from the source, summarised below:

Figure 1.1 SIPs House Construction Site Episkopy Cyprus
Courtesy of SIP Co Ltd.

Figure 1.3.1 Two Storey House Construction
Hugh Morrison (author). SIP Co Ltd project

Figure 1.3.2 Interior Roof (*of* figure 1.3.1)
Hugh Morrison (author). SIP Co Ltd project

Figure 1.3.3 Completed House
Hugh Morrison (author). SIP Co Ltd project

Figure 1.3.4 Interior (*of* figure 1.3.3)
Hugh Morrison (author). SIP Co Ltd project

Figure 1.3.5 SIPs Units Off-site Assembly
Courtesy of New Meaning Foundation (John Evans)

Figure 1.3.6 Modular Accommodation Block
Hugh Morrison (author). SIP Co Ltd project

Figure 1.3.8 School Classroom in Construction
Courtesy of Colin Shea Design Associates

Figure 1.3.9 School Classroom Block
Courtesy of Net Zero Buildings

Figure 1.3.10 Showroom - Office
Courtesy of AF Jones Stonemasons

Figure 1.3.11 Residential Apartment
Hugh Morrison (author) . SIP Co Ltd project

Figure 1.3.12 Port Lockroy Nissen Hut
Courtesy of UK Antarctic Heritage Trust. SIP Co Ltd project

Index

Index

Index

Index

Index

Index

Index

www.ingramcontent.com/pod-product-compliance
Lightning Source LLC
LaVergne TN
LVHW022217200926
848481LV00006B/2002